Electrical Machines

Electrical Machines

Edited by **Jeremy Giamatti**

WILLFORD **P**RESS

New York

Published by Willford Press,
118-35 Queens Blvd., Suite 400,
Forest Hills, NY 11375, USA
www.willfordpress.com

Electrical Machines
Edited by Jeremy Giamatti

© 2016 Willford Press

International Standard Book Number: 978-1-68285-028-2 (Hardback)

Contents

Preface

Over the recent decade, advancements and applications have progressed exponentially. This has led to the increased interest in this field and projects are being conducted to enhance knowledge. The main objective of this book is to present some of the critical challenges and provide insights into possible solutions. This book will answer the varied questions that arise in the field and also provide an increased scope for furthering studies.

Electrical machines like transformers, generators and motors form a significant part of our day to day lives. This adds to the importance of electrical engineering. This book covers in detail some existent theories and innovative concepts revolving around electrical machines like low and high power converters, semiconductors, power devices and components, motor drives, etc. This book is a resource guide for professionals as well as students.

I hope that this book, with its visionary approach, will be a valuable addition and will promote interest among readers. Each of the authors has provided their extraordinary competence in their specific fields by providing different perspectives as they come from diverse nations and regions. I thank them for their contributions.

Editor

Characterization and Modeling of Power Electronics Device

M. N. Tandjaoui, C. Benachaiba, O. abdelkhalek, B. Denai, Y. Mouloudi
University of Bechar, Department of Technology, BP 417 Algeria

Keyword:

GTO
IGBT
MOStransistor
Power converter
Power electronics devices
Semiconductors devices

ABSTRACT

During the three decades spent, the advances of high voltage/current semiconductor technology directly affect the power electronics converter technology and its progress. The developments of power semiconductors led successively to the appearance of the elements such as the Thyristors, and become commercially available. The various semiconductor devices can be classified into the way they can be controlled, uncontrolled category such as the Diode when it's on or off state is controlled by the power circuit, and second category is the fully controlled such as the Metal Oxide Semiconductor Field Effect Transistor (MOSFET), and this category can be included a new hybrid devices such as the Insulated Gate Bipolar Transistor (IGBT), and the Gate Turn-off Thyristor (GTO). This paper describes the characteristics and modeling of several types of power semiconductor devices such as MOSFET, IGBT and GTO.

Corresponding Author:

Abdelkhalek Othmane,
Department of Technology,
University of Bechar,
Bechar University Center BP 417 Bechar 08000, Algeria.
Email: chellali@netscape.net

1. INTRODUCTION

Power electronics is the technology of converting electric power from one from to another using electronic power device. It also refers to a subject of research in electrical engineering which deals with design, control, computation and integration of nonlinear, time varying energy processing electronic systems with fast dynamics. With "classical" electronics, electrical currents and voltage are used to carry information, whereas with power electronics, they carry power. Thus, the main metric of power electronics becomes the efficiency. The capabilities and economy of power electronics system are determined by the active devices that are available. Their characteristics and limitations are a key element in the design of power electronics systems.

Several types of solid state power semiconductor devices have been developed to control of output parameters, such as voltage, current or frequency. In a state power converter, the power semiconductor devices function as switches, which operate statically, that is, without moving contact [1].

A semiconductor material is one whose electrical properties lie in between those of insulators and good conductors. Examples are: germanium and silicon. In terms of energy bands, semiconductors can be defined as those materials which have almost an empty conduction band and almost filled valence band with a very narrow energy gap (of the order of 1 eV) separating the two. Semiconductor may be classified as under:

Semiconductors

Intrinsic or pure
semiconductors

Extrinsic or impure
semiconductors

N-Type   P-Type

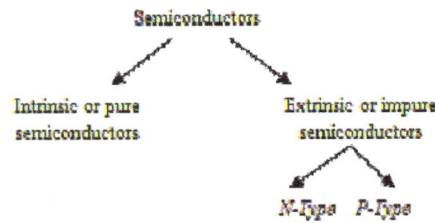

Figure 1. Type of Semiconductors

An intrinsic semiconductor is one which is made of the semiconductor material in its extremely pure form. Alternatively, an intrinsic semiconductor may be defined as one in which the number of conduction electrons is equal to the number of holes.

Those intrinsic semiconductors to which some suitable impurity or doping agent or doping has been added in extremely small amounts (about 1 part in 108) are called extrinsic or impurity semiconductors [4], [5]. Depending on the type of doping material used, extrinsic semiconductors can be sub-divided into two classes, N-type semiconductors and P-type semiconductors.

The first power electronic device started in 1902 with the development of mercury arc rectifier using to convert alternating current (AC) into direct current (DC). In modern systems the conversion is performed with semiconductor switching devices such as diodes, thyristors and transistor, and others hybrid devices beginning in the 1950s such as SCR, MOSFET, GTO, IGBT... The development of power semiconductor devices has always been a driving force for power electronics systems. In the semiconductor industry, we all understand the importance of design, material selection, assembly manufacturing, reliability, and testing in minimizing power packaging failures [4], [2]. For a long time silicon-based power devices have dominated the power electronics and power system applications.

The semiconductor industry has made impressive progress, particularly in communications, health, automotive, computing, consumer, security, and industrial electronics.

Power handling and dissipation of devices is also a critical factor in design. The current rating of a semiconductor device is limited by the heat generated within the dies and the heat developed in the resistance of the interconnecting leads. Semiconductor devices must be designed so that current is evenly distributed within the device across its internal junctions, its forward voltage drop in the conducting state translates into heat that must be dissipated when they require specialized heat sinks or active cooling systems to keep their junction temperature from rising too high. Exotic semiconductors such as silicon carbide have an advantage over straight silicon in this respect, and germanium, once the main-stay of solid-state electronics is now little used due to its unfavorable properties at high temperature.

In the 1980s, the development of power semiconductor devices took an important turn when new process technology was developed that allowed integration of MOS and bipolar junction transistor (BJT) technologies on the same chip. Thus far, two devices using this new technology have been introduced: insulated bipolar transition (IGBT) and MOS controlled thyristor (MCT). Many integrated circuit (IC) processing methods as well as equipment have been adapted for the development of power devices. However, unlike microelectronic ICs, which process information, power device ICs process power and so their packaging and processing techniques are quite different [2], [5].

Several attributes dictate how devices are used. Devices such as diodes conduct when a forward voltage is applied and have no external control of the start of conduction. Power devices such as SCRs and thyristors allow control of the start of conduction, but rely on periodic reversal of current flow to turn them off. Devices such as GTO, IJBT, and MOSFET provide full switching control and can be turned on or off without regard to the current flow through them. The control input characteristics of a device also greatly affect design; sometimes the control input is at a very high voltage with respect to ground and must be driven by an isolated source. As efficiency is at a premium in a power electronic converter, the losses that a power electronic device generates should be as low as possible.

Even though the technologies with silicon (Si)-based power devices are mature, inherent material restrictions limit their performance in high voltage, high power, high switching frequency and high temperature applications. Bipolar power devices, such as insulated-gate bipolar transistors (IGBTs), can handle high power, but the switching speed is limited by the devices' structure. Unipolar power devices, like metal-oxide semiconductor field effect transistors (MOSFETs), can be switched at high frequency, but suffer from relatively high on-state resistance. Furthermore, Silicon power devices generally can only withstand operational temperature of 150°C and can require a substantial cooling system [3].

A gate turn-off thyristor (known as a GTO) is a three-terminal power semiconductor device that belongs to a thyristor family with a four-layer structure. They also belong to a group of power semiconductor devices that have the ability to fully control on and off states via the control terminal (gate). The design, development, and operation of the GTO are easier to understand if we compare it to the conventional thyristor. Like a conventional thyristor, applying a positive gate signal to its gate terminal can turn on a GTO. Unlike a standard thyristor, a GTO is designed to turn off by applying a negative gate signal [2], [4], [7].

The aim of this paper is to present the principles underlying power conversion by the use of static switches and the techniques employed for controlling output parameters such as voltage, current, power, frequency and waveform. We shall present a characteristic and modeling of two types of power electronics devices such as IGBT and GTO and its results of simulation when we used the MATLAB/SIMULINK to simulate these devices. In a progressive sequence, we shall present all the important types of power converters that have proved useful in the application areas of electric power. We shall also present important application areas, and this will bring out how converter schemes and control strategies can be tailored to meet specific needs.

2. CHARACTERIZATION AND MODELING OF POWER ELECTRONICS DEVICES

The controlled valves, IGBT and GTO, take a significant place in converters of all kinds. As these new types of valves can as well cut conduction as to start it, the IGBT and the GTO gradually replace the thyristors in the applications where one was formerly to call upon forced commutation.

2.1. GTO Thyristor Characteristic

This semiconductor device, as the name implies, is a hybrid device that behaves like a thyristor. However, it has an added feature that the provide gate control allows the designer to turn the device on and off if and when desired [ebook_Power Electronic Control in Electrical Systems]. The symbol for a GTO and i-v characteristics are plotted in figure 1 (a) and (b) respectively.

A high degree of interdigitation is required in GTOs in order to achieve efficient turn-off. The most common design employs the cathode area separated into multiple segments (cathode fingers) and arranged in concentric rings around the device center [2].

Apart from their ability to interrupt the anode current by injecting a current in the gate, the GTO are very similar to thyristors. The conduction of GTO is initiated by injecting a positive current in the trigger. To maintain conduction in the GTO, the anode current must not fall below a threshold called holding current. The anode current is blocked by injecting a current in the trigger substantial negative for a few microseconds. In order to ensure the locking, the current injected into the trigger must be about one-third of the current flowing in the anode. The GTO switches are therefore of great power, which can control the currents of a few thousand amps at voltages up to 4000v [7].

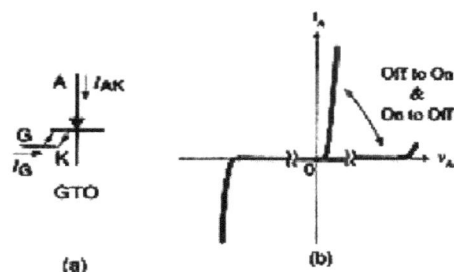

Figure 2. GTO; (a) circuit symbol; (b) i-v characteristics

It should be noted that although the GTO can be turned on like a thyristor, with a low positive gate current pulse, a large negative pulse is required to turn it off. These are relatively slow devices when compared with other fully controlled semiconductors.

The maximum switching frequency attainable is in the order of 1Khz. The voltage and current rating of the commercially available GTOs are compared to the thyristors approaching 6.5Kv, 4.5KA and are expected to increase to cover completely the area occupied by thyristor as [5].

A major variation on the thyristor is the GTO (Gate Turn-Off Thyristor). This is a thyristor where the structure has been tailored to give better speed by techniques such as accurate lifetime killing, fine finger or cell structures and "anode shorts" (short circuiting P+ and N- at the back in order to decrease the current gain of the PNP transistor).As a result, the product of the gain of both NPN and PNP is just sufficient to keep the GTO conductive. A negative gate current is enough to sink the whole current from the PNP and turn the device off.

A GTO shows much improved switching behavior but still has the tail as described above. Lower power applications, especially resonant systems, are particularly attractive for the GTO because the turn-off losses are virtually zero [6], [7].

2.2. IGBT Characteristic

The IGBT is a hybrid semiconductor device that literally combines of the MOSFET and BJT. The MOSFET is a transistor device capable of switching fast with low switching losses. It cannot handle high power and is mostly suited for low-power application. Thus, the BJT is so named because the conduction is due to the movement of electrons and holes within the transistor.

Specifically, it has the switching characteristics of the MOSFET with the power handling capabilities of the BJT. It is a voltage-controlled device like the MOSFET but has lower conduction losses. Furthermore, it is available with higher voltage and current ratings. There are a number of circuit symbols for the IGBT with the most popular and the typical i-v characteristics are plotted in Figure 4.

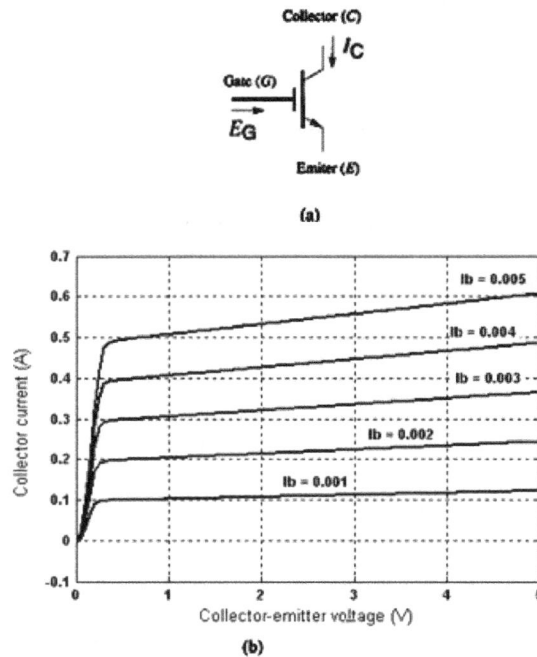

Figure 3. IGBT features; (a) circuit symbol; (b) i-v characteristics

The IGBTs are faster switching devices than the BJTs but not as fast as the MOSFETs. The IGBTs have lower on-state voltage drop even when the blocking voltage is high. The IGBT is the most popular device for AC and DC motor drives reaching power levels of a few hundred Kw. It has also started to make its way in the high voltage converter technology for power system application [5], [7].

The IGBT is a transistor whose conduction is primed and unprimed by applying an appropriate voltage on the trigger (the base). As in a conventional transistor, the three terminals are named C collector, emitter E and base B.

The IGBT can withstand much higher currents the currents ID of the MOSFET. Therefore, they can order higher powers. Compared with the GTO, the BJT, MOSFET and IGBT can initiate and interrupt the flow of anode current with greater speed. This allows these semiconductors to operate at much higher frequencies. This results in a reduction in the size, weight and cost of devices using these valves [7].

3. SEMICONDUCTOR SWITCHING-POWER PERFORMANCE

The power frequency ranges of the various semiconductors discussed in the previous sections are summarized in Figure 3. It is clear that the thyristor dominates the ultra-high power region for relatively low frequencies.

The GTO is the next device when it comes to power handling capabilities extending to frequencies of a few hundred Hz. The IGBT occupies the area of medium power with the ability to operate at relatively higher frequencies.

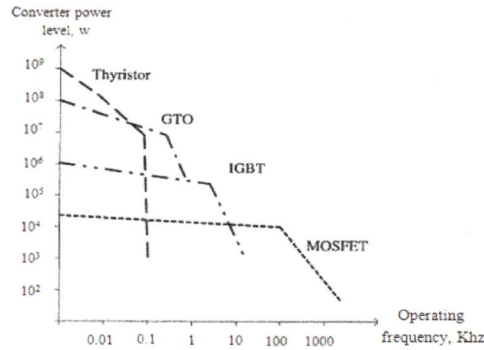

Figure 4. Power Converter Level and Frequency for Various Semiconductor Devices

The tendency over the next few years is to have the GTO extend its power area towards the thyristor level. At the same time, the IGBT will also extend its power ability towards the GTO with higher switching frequency [5].

4. MATLAB SIMULATIONS RESULTS

An analog simulation of power electronics devices such as MOSFET, IGBT and GTO is proposed in order to evaluate the behavior of this component in its electrical environment. In the first of this paper, the principal characteristics of GTO and IGBT are presented.

The Figure 5 showing the following out waveforms voltage's of supply, MOSFET, IGBT, and GTO. The setting with the state on is mainly subjected to the constraint of the blocking of the diode positioned. This is related to the reverse current of diode.

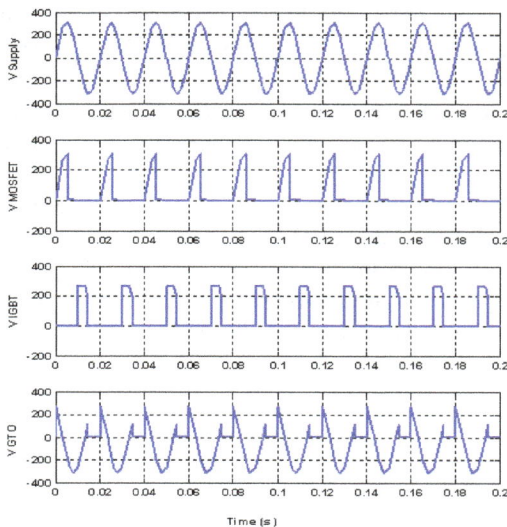

Figure 5. The Voltage Waveforms of Supply, MOSFET, IGBT and GTO

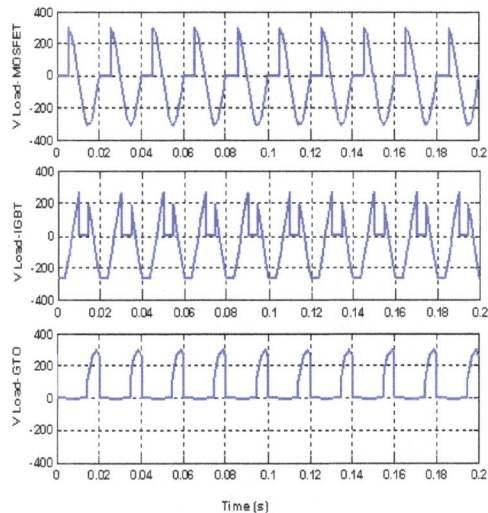

Figure 6. The Voltage Waveforms Loads with Devices MOSFET, IGBT and GTO

But the turn-on is also depend to the possible disturbances in the grid control, and just like for the phase of turn-off the behavior of the monolithic structures to closing is influenced by the impact of the circuit surrounding them.

In this all test of simulation for different components devices, the load is proposed as a resistance of 50Ω. In Figure 6, we can see the following forms of voltage of load when it is operated for these different electronics devices.

The waveforms of current in output of these devices can be observed in Figure 7 and show that the current fluctuation with the setting the state on component is quite present. However, the quantity of current is depended to the current of covering of the diode. The waveforms of load currents are presented in Figure 8 are similar as these presented in figure7 because, in this case of study, we used a purely resistive load.

Finally, let us conclude from it that these electronics devices are major pollutants in the power quality, that us obliges taken into account their use in order to preserve the good quality of the power supply.

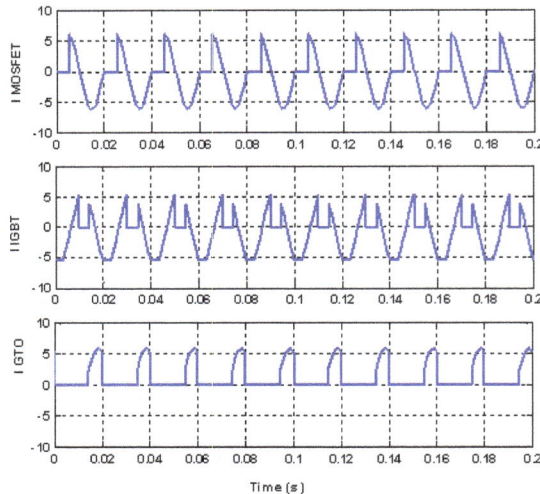

Figure7. The Current Waveforms Output of MOSFET, IGBT and GTO

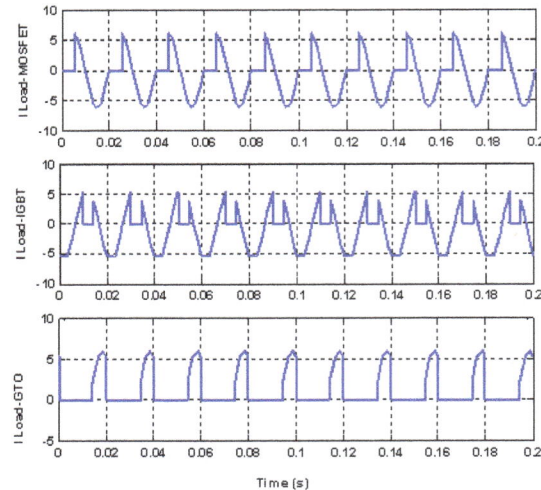

Figure 8. The Current Waveforms Loads Used MOSFET, IGBT and GTO

5.   CONCLUSION

The electrical characterization of several active switches as MOSFET, IGBT and GTO was presented in this paper. The behavioral of these devices developed using MATLAB / SIMULINK when the power electronic circuits are simulated before they are produced to test how the circuits respond under certain conditions in electrical aims application.

The study theoretical approach of these electronic power component characteristics was presented. It is noteworthy that the behavior enables also to predict certain behaviors of this electronic power component in the electrical circuit.

Finally, the simulation of these electronics devices in real time permits to extract the state of the components is the application of electronics for the control and conversion of electric power. Applications of power electronics range in size from a switched mode power supply in an AC adapter, like FACTS (Flexible alternative current transmission systems) and DC motor drives used to operate pumps and to interconnect electrical grids with a novel technique of transmission based on power electronic like HVDC (high voltage direct current systems).

REFERENCES
[1] Batarseh I. *Power electronic circuits*. John Wiley. 2004. ISBN 0-471-12662-4, 9780471126621.
[2] MH RASHID. Gate Turn-Off Thyristors', Electrical and Computer Engineering, University of West Florida 11000 University Parkway, Pensacola, Florida 32514-5754, USA.
[3] Y Cui, M Chinthavali F, Xu L. Tolbert. Characterization and Modeling of Silicon Carbide Power Devices and Paralleling Operation. IEEE xplore. 2012.
[4] F Mazda. Power Electronics Handbook, Third edition. Newnes, ISBN 0 7506 2926 6, 2003.

[5]　E Acha, VG Agelidis, OA Lara, TJE Miller. Power electronic control in electrical systems' Newnes power engineering series. 2002; ISBN 0750651261.

[6]　T Wildi, G Sybille. Electrotechnique. Paris, France. ISBN PUL 2-7637-8185-3. 2005.

2

Photovoltaic Cell Fed 3-Phase Induction Motor Using MPPT Technique

Gudimetla Ramesh*, Kari Vasavi**, S.Lakshmi Sirisha*
* Departement of Electrical and Electronics Engineering, Jawaharlal Nehru technological university Kakinada
** Departement of Electrical and Electronics Engineering, K L University

ABSTRACT

Keyword:

DC-DC Converter
Dspic4011
MPPT
PhotoVoltaic

This Paper emphasizes on proposing a cost effective photovoltaic (PV) fed 3 phase Induction motor drive which serves for rural pumping applications. Generally in a standalone system, the PV unit will charge the battery and the battery set up in turn will serve as a source for the inverter. A new single stage battery less power conversion is employed by designing a maximum power point tracker (MPPT) embedded boost converter which makes the overall cost of the setup to go down considerably. The realized as a prototype consisting PV array of 500watts, MPPT aided boost converter, three phase inverter and a three phase squirrel cage induction drive of 300 watts. An efficient and low cost micro controller dspic4011 is used a platform to code and implement the prominent perturb and observe MPPT technique. Sinusoidal pulse width modulation (SPWM) is the control technique employed for the three phase inverter. To validate the experimental results simulation of the whole set up is carried out in matlab /simulink environment. Simulation and hardware results reveal that the system is versatile.

Corresponding Author:

S.Lakshmi Sirisha,
Departement of Electrical and Electronics Engineering,
Jawaharlal Nehru technological university Kakinada.
Email: slsirisha.s@gmail.com

1. INTRODUCTION

The induction motors do not require an electrical connection between stationary and rotating parts of the motor. Therefore, they do not need any mechanical commutator (brushes), leading to the fact that they are maintenance free motors. Induction motors also have low weight and inertia, high efficiency and a high overload capability. Therefore, they are cheaper and more robust, and less proves to any failure at high speeds. A variable frequency is required because the rotor speed depends on the speed of the rotating magnetic field provided by the stator. A variable voltage is required because the motor impedance reduces at low frequencies and consequently the current has to be limited by means of reducing the supply voltages. Before the days of power electronics, a limited speed control of induction motor was achieved by switching the three-stator windings from delta connection to star connection, allowing the voltage at the motor windings to be reduced. Induction motors are also available with more than three stator windings to allow a change of the number of pole pairs.

2. PHOTOVOLTAIC TECHNOLOGY

Converting the sun's radiation directly into electricity is done by solar cells. These cells are made of semiconducting materials similar to those used in computer chips. When sunlight is absorbed by these materials, the solar energy knocks electrons loose from their atoms, allowing the electrons to flow through the material to produce electricity. This process of converting light (photons) to electricity (voltage) is called the photovoltaic effect. Photovoltaic's (PV) are thus the field of technology and research related to the application of solar cells that convert sunlight directly into electricity. Solar cells, which were originally developed for space applications in the 1950s, are used in consumer products such as calculators or watches, mounted on roofs of houses or assembled into large power stations. Today, the majority of photovoltaic modules are used for grid-connected power generation, but a smaller market for off-grid power is growing in remote areas and developing countries. Given the enormous potential of solar energy, photovoltaic may well become a major source of clean electricity in the future. However, for this to happen, the electricity generation costs for PV systems need to be reduced and the efficiency of converting sunlight into electricity needs to increase.

3. PHOTOVOLTAIC ARRAYS

Due to the low voltage of an individual solar cell typically 0.5V, several cells are wired in series in the manufacture of a "laminate". The laminate is assembled into a protective weatherproof enclosure, thus making a photovoltaic module or solar panel. These solar panels are linked together to form photovoltaic Arrays. The panels are connected in series The current through the cell is constant and the voltage across the cell adds up. The panels are connected in parallel The voltage through the cell is constant and the current across the cell adds up.

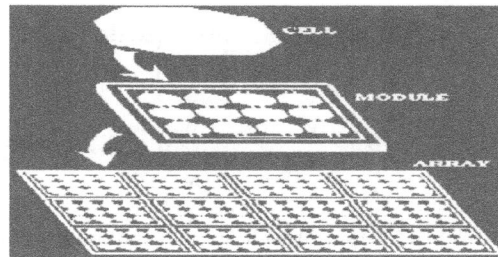

Figure 1. PV array

When photons of light strike the material, however, some normally non-mobile electrons in the material absorb the photons, and become mobile by virtue of their increased energy. This creates new holes too - which are just the vacancies created by the newly created mobile electrons. Because of the "built in" electric field, the new mobile electrons in the n-material cannot cross over into the p-material. In fact, if they are created near or in the junction where the electric field exists, they are pushed by the field towards the upper surface of the n-material. If a wire is connected from the n-material to the p-material, however, they can flow through the wire, and deliver their energy to a load.

On the other hand, the holes created in the n-material, which are positively charged, are pushed over into the p-material. In fact, what is really happening here is that an electron from the p-material, which was also made mobile by the adsorption of a photon, is pushed by the electric field across the junction and into the n-material to fill the newly created hole. This completes the circuit as the electrons flows in all the ways around the circuit, dropping the energy they acquired from photons at a load.

4. BLOCK DIAGRAM OF PHOTOVOLTAIC PUMPING SYSTEM

The design of an effective PV pumping system without the use of a battery bank represents a significant challenge. It is necessary to deal with the effect of the stochastic nature of solar installation on the entire energy conversion chain, including the nonlinear characteristics of PV pumping, the voltage boost converter, and the electromechanical power conversion device. In general terms, it is necessary to obtain the best performance from each system component over a wide input power range.

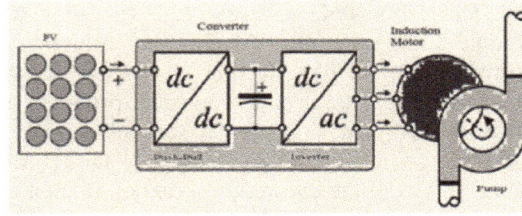

Figure 2. Photovoltaic pumping systems

Currently, solar water pumps are used in the western United States as well as in many other countries or regions with abundant sunlight. Solar pumps have proven to be a cost effective and dependable method for providing water in situations where water resources are spread over long distances, power lines are few or non-existent, and fuel and maintenance costs are considerable.

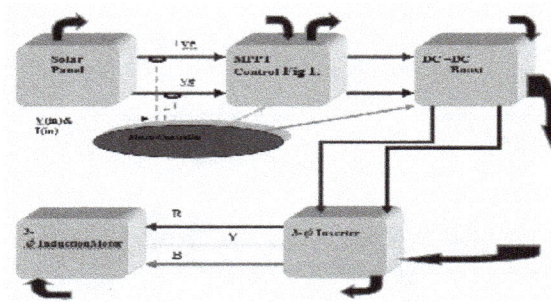

Figure 3. MPPT control technique of Induction Motor

5. PV CELL MODELLING

Renewable energies are on demand these days and amongst them, Solar is the most popular because of the basic fact that it is clean, green and its availability is boundless. It's almost been decades and using solar energy has become a tradition now and the best part is that it is still escalating at tremendous pace with newer and more efficient technologies. Right from monocrystalline and polycrystalline we have leaped towards amorphous silicon, thin film technology, Cd-teThin film, CIGS thin film and flexible thin films. Though these trends are still in the research stag and not yet popular two players who have contributed to almost 90% of the solar panels are Monocrystalline and Polycrystalline Silicon. Solar cells are connected in series and parallel to get the desired output as a single solar cell could only contribute a peak voltage of 0.5 to 0.7 volt. Such designed unit is called a PV panel and these panels are in turn arranged series and parallel to form PV array.

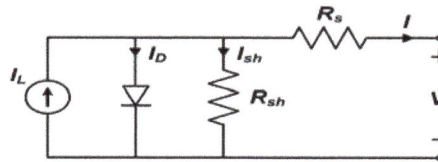

Figure 4. PV cell equivalent diagram

In the equivalent circuit, I_{pv} and V_{pv} are the PV current and voltage respectively. R_s and R_{sh} are the series and shunt resistances respectively. Now the current to the load is given by:

$$I = N_p I_{pv} - N_p I_s \left[\exp\left[\frac{q(V + R_s I)}{nN_s KT} - 1\right]\right] - \frac{V + R_s I}{R_{sh}} \tag{1}$$

In this equation, I_{pv} is the photocurrent, I_s is the reverse saturation current of the diode, q is the electron charge, V is the voltage across the diode, K is the Boltzmann's constant, T is the junction temperature, n is the ideality factor of the diode, and R_s and R_{sh} are the series and shunt resistors of the cell, respectively. N_s and N_p are the number of cells connected in series and parallel respectively.

As mentioned earlier PV current is a function of temperature and solar irradiation.

$$I_{pv} = [I_{sc} + K_i(T - 298)]\frac{\beta}{1000} \tag{2}$$

Where K_i=0.0017 A/∘C is the cell's short circuit current temperature coefficient and β is the solar radiation (W/m2). The diode reverse saturation current varies as a cubic function of the temperature and it can be expressed as:

$$I_s(T) = I_s\left[\frac{T}{T_{nom}}\right]^3 \exp\left[\left[\frac{T}{T_{nom}} - 1\right]\frac{E_g}{nV_t}\right] \tag{3}$$

Where I_s is the diode reverse saturation current, T_{nom} is the nominal temperature, E_g is the band gap energy of the semiconductor and V_t is the thermal voltage.

6. PERTURB AND OBSERVE MPPT TECHNIQUE:

The basic disadvantages which PV system face is that the irradiance of sun is never constant and hence it is difficult to yield the full performance from the panel. Hence whenever the source is varying one it is often better to work on the existing output obtained and mould it accordingly such that despite the loss at input side, the output remains unaltered. This is what a MPPT does precisely. Maximum Power Point Tracking is electronic tracking-usually digital. The charge controller looks at the output of the panels, and compares it to the battery voltage. It then figures out what is the best power that the panel can put to charge the battery. It takes this and converts it to best voltage to get maximum AMPS to the battery. MPPTs are most effective when weather is cold, when the battery charge is lo/w and the cable wires used for connection are long. Hence these days MPPTs have become mandatory. In perturb and observe technique, the controller adjusts the voltage by a small amount from the array and measures power; if the power increases, further adjustments in that direction are tried until power no longer increases. This is called the perturb and observe method and is most common, although this method can result in oscillations of power.

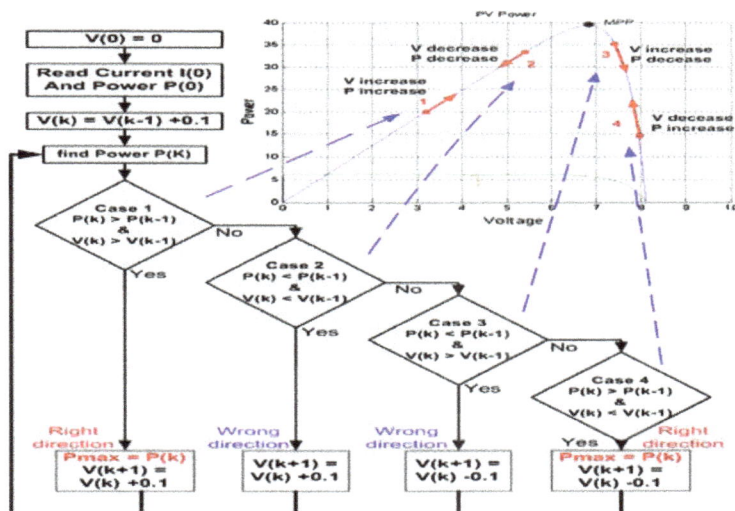

Figure 5. Perturb and observe MPPT technique

7. INDUCTION MOTOR:

The only effective way of producing an infinitely variable induction motor speed drive is to supply the induction motor with three phase voltages of variable frequency and variable amplitude. A variable frequency is required because the rotor speed depends on the speed of the rotating magnetic field provided by the stator. A variable voltage is required because the motor impedance reduces at low frequencies and consequently the current has to be limited by means of reducing the supply voltages. Before the days of power electronics, a limited speed control of induction motor was achieved by switching the three-stator windings from delta connection to star connection, allowing the voltage at the motor windings to be reduced. Induction motors are also available with more than three stator windings to allow a change of the number of pole pairs. However, motor with several windings is more expensive because more than three connections to the motor are needed and only certain discrete speeds are available. Another alternative method of speed control can be realized by means of a wound rotor induction motor, where the rotor winding ends are brought out to slip rings.

8. INDUCTION MOTOR SPECIFICATIONS:

Table 1. Induction Motor specifications

Rated Power	0.37kW/0.5 HP
Rated Current	1.4A
Voltage	415V
Speed	1330 rpm
% Efficiency	64
Frequency	50Hz

9. RESULTS

Computer simulation is a widely accepted tool for analysis and design of electrical systems, the large interconnected power systems. Digital simulation tools like MATLAB offer a convenient mechanism to solve these problems.

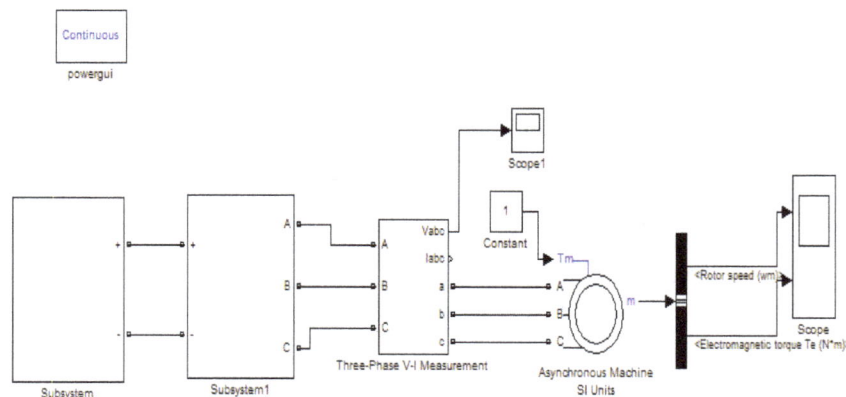

Figure 6. Complete system simulation Diagram of PV Cell Fed induction Motor

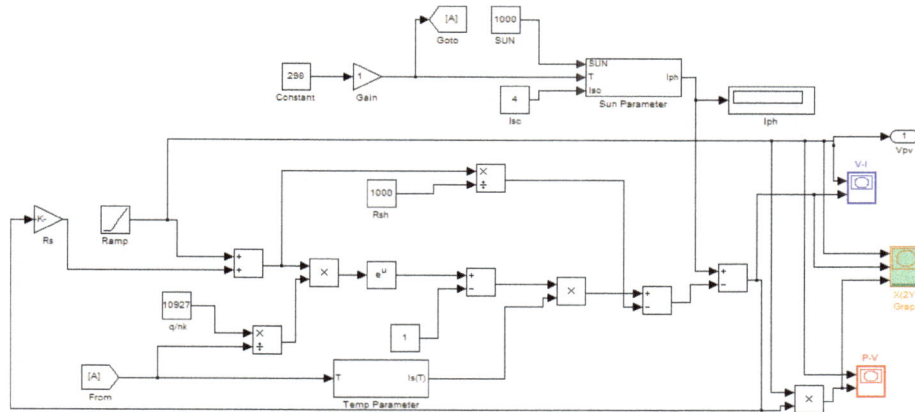

Figure 7. Subsystem of PV Cell

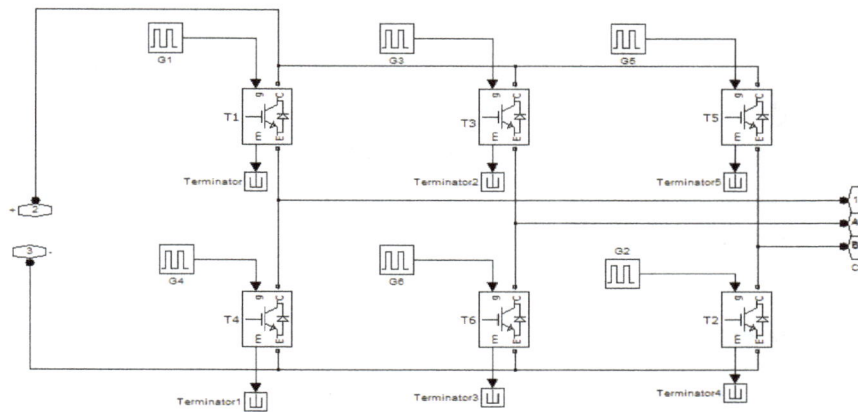

Figure 8. Subsystem of Voltage Sourse Inverter

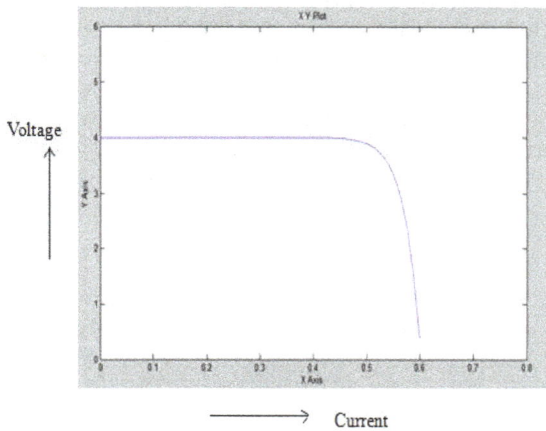

Figure 9. V-I Charactorstics Of Induction Motor

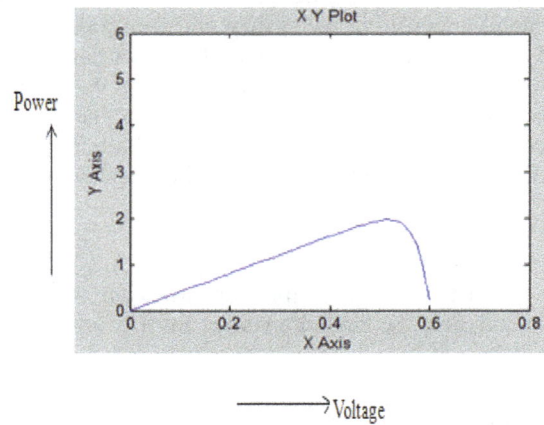

Figure 10. P-V Charactrostics of Induction Motor

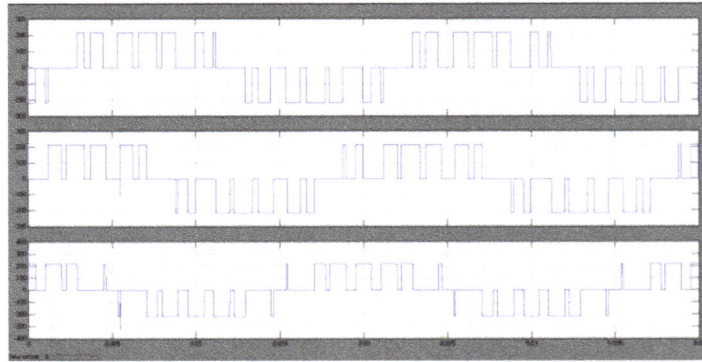

Figure 11. Inverter Output Voltages of Va, Vb, Vc

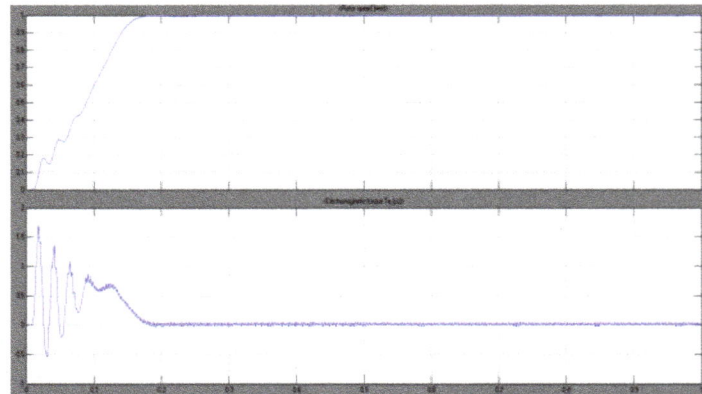

Figure 12. Speed and Torque characteristics Induction motor

10. CONCLUSION

The main objectives were to achieve maximum power output from the PV array and to inject a high quality AC current into the grid to transfer that power. To that aim, the PV cell equivalent circuit model was obtained to construct the system and then focus was directed towards the power conditioning system (PCS) and its controls. The first stage of the PCS was a DC-DC boost converter responsible for extracting maximum power from the PV array and increasing its output voltage. The second stage of the PCS was a current controlled voltage source inverter (VSI) that converted the DC power of the array into AC power and injected it into the grid. The control technique relied on transforming the three phase currents and voltages into a rotating reference frame and then regulated the resulting *dq* current components.

ACKNOWLEDGEMENTS

The authors gratefully acknowledge the K.Vasavi, K L University, Vijayawada. and particularly the Principal, Prof. K. Phanidra Kumar for the financialy support and for the facilities offered in these researches.

REFERENCES

[1] X Wang, Z Fang, J Li, L Wang, S Ni. Modeling and control of dual-stage high-power multifunctional PV system in *d–q–o* coordinate. *IEEE Trans. Ind. Electron.*, 2013; 60(4): 1556–1570.
[2] B Indu Rani, G Saravana Ilango, C Nagamani. Control strategy for power flow management in a PV system supplying DC loads. *IEEE Trans. Ind. Electron.*, 2013; 60(8): 3185–3194.
[3] N Femia, G Petrone, G Spagnuolo, M Vitelli. A technique for improving P&O MPPT performances of double-stage grid-connected photovoltaic systems. *IEEE Trans. Ind. Electron.*, 2009; 56(11): 4473–4482.
[4] ES Sreeraj, K Chatterjee, S Bandyopadhyay. One cycle controlled single-stage, single-phase voltage sensor-less grid-connected PV system. *IEEE Trans. Ind. Electron.*, 2013; 60(3): 1216–1224.
[5] AI Bratcu, I Munteanu, S Bacha, D Picault, B Raison. Cascaded DC–DC converter photovoltaic systems: Power optimization issues. *IEEE Trans. Ind. Electron.*, 2011; 58(2): 403–411.

[6] G Petrone, G Spagnuolo, M Vitelli. A multivariable perturband observe maximum power point tracking technique applied to a singlestage photovoltaic inverter. *IEEE Trans. Ind. Electron.*, 2011; 58(1): 76–84.

[7] W Li, X He. Review of nonisolated high-step-up DC/DC converters in photovoltaic grid-connected applications. *IEEE Trans. Ind. Electron.*, 2011; 58(4): 1239–1250.

[8] R Kadri, J Gaubert, G Champenois. An improved maximum power point tracking for photovoltaic grid-connected inverter based on voltageoriented control. *IEEE Trans. Ind. Electron.*, 2011; 58(1): 66–75.

[9] G Petrone, G Spagnuolo, M Vitelli. An analog technique for distributed MPPT PV applications. *IEEE Trans. Ind. Electron.*, 2012; 59(12): 4713–4722.

[10] J Chavarria, D Biel, F Guinjoan, C Meza, J Negroni. Energybalance control of PV cascaded multilevel grid-connected inverters for phase-shifted and level-shifted pulse-width modulations. *IEEE Trans. Ind. Electron.*, 2013; 60(1): 98–111.

3

Comparison Analysis of Indirect FOC Induction Motor Drive using PI, Anti-Windup and Pre Filter Schemes

M.H.N Talib*, Z. Ibrahim**, N. Abd. Rahim***, A.S.A. Hasim****
*,** Departement of Electrical Engineering, Universiti Teknikal Malaysia Melaka
*** UMPEDAC, Universiti Malaya, Kuala Lumpur, Malaysia
**** Faculty of Engineering, Universiti Pertahanan Nasional Malaysia, Kuala Lumpur, Malay

Keyword:

Field Oriented Control (FOC)
SVPWM
Induction Motor Drive
PI controller
Speed Control

ABSTRACT

This paper presents the speed performance analysis of indirect Field Oriented Control (FOC) induction motor drive by applying Proportional Integral (PI) controller, PI with Anti-Windup (PIAW) and Pre- Filter (PF). The objective of this experiment is to have quantitative comparison between the controller strategies towards the performance of the motor in term of speed tracking and load rejection capability in low, medium and rated speed operation. In the first part, PI controller is applied to the FOC induction motor drive which the gain is obtained based on determined Induction Motor (IM) motor parameters. Secondly an AWPI strategy is added to the outer loop and finally, PF is added to the system. The Space Vector Pulse Width Modulation (SVPWM) technique is used to control the voltage source inverter and complete vector control scheme of the IM drive is tested by using a DSpace 1103 controller board. The analysis of the results shows that, the PI and AWPI controller schemes produce similar performance at low speed operation. However, for the medium and rated speed operation the AWPI scheme shown significant improvement in reducing the overshoot problem and improving the setting time. The PF scheme on the other hand, produces a slower speed and torque response for all tested speed operation. All schemes show similar performance for load disturbance rejection capability.

Corresponding Author:

M.H.N Talib,
Faculty of Electrical Engineering,
Universiti Teknikal Malaysia Melaka,
Hang Tuah Jaya, 76100 Durian Tunggal, Melaka, Malaysia.
Email: hairulnizam@utem.edu

1. INTRODUCTION

Vector control or field oriented control (FOC) drive is one of the most popular choices of variable speed drive application industries. Since the advent of indirect FOC in 70's, the proportional integral (PI) controller scheme has been widely used in variable speed drive motor. However, there are several types of controller scheme such as PI control, fuzzy logic control, artificial intelligent control and variable structure controlled which can be utilized to get the best performance of the motor [1]-[7]. The main reason PI controller is well accepted is due to the simple structure which can be easily understood and implemented. This technique is able to independently control the torque and the flux-producing component of the stator current in a wide speed range.

However, in order to ensure the PI controller to work efficiently, the value of proportional gain (Kp) and integral gain (Ki) must be tuned correctly. The performance of the motor really depends on the gain of the PI controllers. However, in most cases, these gains are determined by a trial and error tuning technique which requires practical experience and may lead to time consumption. Even though, there are numbers of tuning technique such as Ziegler-Nichols methods and first order plus time delay method, certain knowledge of process control is required and even that will not ensure the best control performance [2], [3], [8]-[10]. On the other hand, the general second order method offers simpler technique and more mathematical formulation approached method. Thus, this method had been applied in getting all the PI values for the analysis in this paper.

Indirect FOC method itself faced a problem with parameter variation caused by the motor heating phenomenon and saturation [11]. This variation causes detuning problem in the decoupling operation and produce errors in the motor output values. Thus, a robust controller designed is necessary to adapt with the parameter variations and decoupling operation. In addition, it able to produce robust solution by applying the integral of time multiplied by the absolute of the error (ITAE) criterion method [2], [10]. Conventional or linear PI controller does not have output magnitude limiters, which could cause damage to the real system due to relatively large output value. Introducing integrator limiter and saturation limiter provide some protection to the system. However, this saturation limiter accumulates error, thus producing large overshoot, slow settling time and sometimes instability to the system [3], [4], [12]-[15]. Thus, PI controller with anti wind up was introduced. There are several Anti-Windup PI controllers to solve this wind up phenomenon such as AWPI with dead zone, AWPI condition, AWPI with tracking and many more. Most of the papers discuss on the anti wind up scheme in solving wind up phenomenon issue and its improvement. Based on the comparative study on the anti windup strategies, the AWPI condition technique found to be the most suitable for usual application due to the performance results, simple structure and less parameter controlled[3]-[4], [12]-[13], [16]. Most of the papers discussed only on the PI and Anti-windup performance at rated speed range. In this paper, the PF analysis is added in the analysis in various speed range demands. The pre filter scheme is able to get rid the unwanted zero in the closed loop system [3], [10].

In this project, the PI controller design is adopted based on the second order system design which has a simpler technique and direct mathematical formulation in comparison to the classical gain tuning method or symmetric optimum criterion[3], [9]-[10], [17]. The performance results of motor behaviors under wide speed range operation and load disturbance are analyzed based on the PI, Anti-Windup and Pre-Filter techniques. As far as the authors' knowledge, no work has been reported on analyzing the speed control motor performance based on this three control techniques together in different speed demand ranged quantitatively.

2. INDIRECT FIELD ORIENTED CONTROL DRIVE

The FOC imitates the concept of separately excited dc motor drive. Through this concept, the torque and the flux are controlled by two independent orthogonal variables known as the armature and field currents. Figure 1 shows the block diagram of indirect FOC scheme.

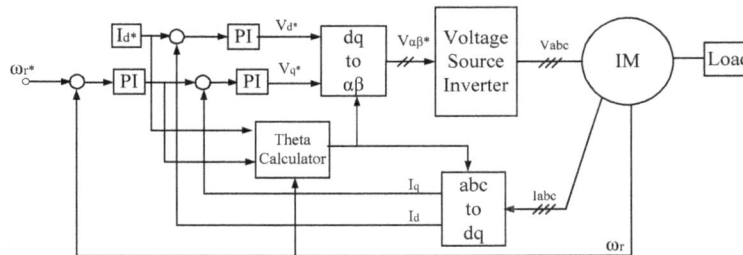

Figure 1. Indirect FOC block diagram

By applying space vector transformation to a three-phase system, the dynamic behavior of induction motor can be represented in mathematical equations as in (1)-(4) in synchronous rotating frame [18], [19].

Stator voltage equations:

$$\bar{V}_{ds} = R_s \bar{I}_{ds} + \frac{d\bar{\varphi}_{ds}}{dt} - \omega_s \bar{\varphi}_{qs}$$

$$\bar{V}_{qs} = R_s \bar{I}_{qs} + \frac{d\bar{\varphi}_{qs}}{dt} + \omega_s \bar{\varphi}_{ds} \tag{1}$$

Rotor Voltage equations:

$$\bar{V}_{dr} = 0 = R_r\bar{I}_{dr} + \frac{d\bar{\varphi}_{dr}}{dt} - (\omega_s - \omega_r)\bar{\varphi}_{qr}$$

$$\bar{V}_{qr} = 0 = R_r\bar{I}_{qr} + \frac{d\bar{\varphi}_{qr}}{dt} + (\omega_s - \omega_r)\bar{\varphi}_{dr} \tag{2}$$

Stator Flux equations:

$$\bar{\varphi}_{ds} = L_s\bar{I}_{ds} + L_m\bar{I}_{dr}$$

$$\bar{\varphi}_{qs} = L_s\bar{I}_{qs} + L_m\bar{I}_{qr} \tag{3}$$

Rotor Flux equations:

$$\bar{\varphi}_{dr} = L_m\bar{I}_{ds} + L_r\bar{I}_{dr}$$

$$\bar{\varphi}_{qr} = L_m\bar{I}_{qs} + L_r\bar{I}_{qr} \tag{4}$$

Where $\bar{V}, \bar{I}, \bar{\varphi}$, are the voltages, current and flux. Meanwhile subscript d, q represent the dq axis while s and r represent stator and rotor component. The stator and rotor resistance and inductance are denoted as Rs, Rr and Ls, Lr, whereas Lm is the mutual inductance. ωs and ωr represent the synchronous speed and mechanical speed respectively.

In the space vector approached, the electromagnetic torque, Te produced by the motor can be expressed in terms of flux and current as follows;

$$T_e = \frac{3}{2}\frac{P}{2}(\bar{\varphi}_{ds}\bar{I}_{qs} - \bar{\varphi}_{qs}\bar{I}_{ds})$$

$$T_e - T_L = J\frac{d\omega_r}{dt} + B\omega_r \tag{5}$$

Where P, T_L, J and B denote the number of poles, external load, inertia and friction of the IM coupled with the permanent magnet dc-machine respectively.

In this system, the rotating coordinate reference frame having direct axis is aligned with the rotor flux vector that rotates at the stator frequency. If the q-component of the rotor flux is assume zero and the electromagnetic torque expression becomes:

$$T_e = \frac{3}{2}\frac{P}{2}\frac{L_m^2}{L_r}\bar{I}_{sd}\bar{I}_{sq} \tag{6}$$

Based on the rotor voltage quadrature axis equation of IM, the rotor flux linkage can be estimated using this formula;

$$\hat{\psi}_r = \frac{L_m I_{ds}}{1 + \tau_r s} \tag{7}$$

Where, τ_r is the rotor time constant.

The slip frequency ω_{sl} is obtained from the rotor voltage direct axis equation by:

$$\omega_{sl} = \frac{L_m R_r I_{qs}}{\hat{\psi}_r L_r} \tag{8}$$

The rotor flux position, θ_e for coordinate transform is generated from the integration of rotor speed, ω_r and slip frequency, ω_{sl}.

$$\theta_e = \int \omega_r + \omega_{sl} \tag{9}$$

The FOC is composed of two inner current loops for flux and torque control. The outer speed loop is cascaded with the torque current loop. The output of this current loop regulate are transformed into stationary reference frame voltage by dq to αβ transformation. Then, these reference voltages are fed to SVPWM modulation process to generate pulse with modulation signal for inverter.

3.   CONTROLLER DESIGN

Based on the mathematical model of the three phase IM, all the current loop and speed of PI controller are calculated by using a second order system for a step input. All the values for proportional (Kp) and integral (Ki) gains of the three PI controllers are determined by comparing the general second order system with the close loop block diagram transfer function.

3.1. PI Controller Scheme

Based on the motor Equation (1), in synchronous reference frame the block diagram of torque and flux component loop can be simplified as in Figure 2 and Figure 3.

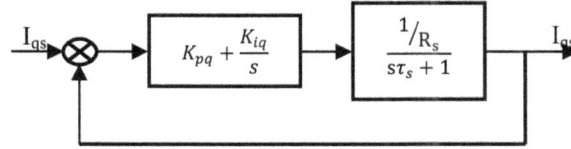

Figure 2. Simplified torque component current loop control

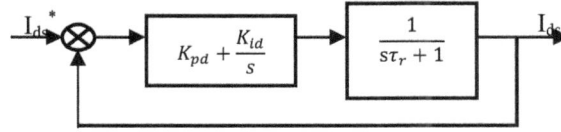

Figure 3. Simplified flux component current loop control

The closed loops equations for torque and flux component above are shown in Equation (10) and (11).

$$\frac{I_{qs}^*}{I_{qs}} = \frac{\left(K_{pq} + \frac{K_{iq}}{s}\right)\left(\frac{1/R_s}{s\tau_s + 1}\right)}{1 + \left(K_{pq} + \frac{K_{iq}}{s}\right)\left(\frac{1/R_s}{s\tau_s + 1}\right)} \tag{10}$$

$$\frac{I_{ds}^*}{I_{ds}} = \frac{\left(K_{pd} + \frac{K_{id}}{s}\right)\left(\frac{1}{s\tau_r + 1}\right)}{1 + \left(K_{pd} + \frac{K_{id}}{s}\right)\left(\frac{1}{s\tau_r + 1}\right)} \tag{11}$$

Where $\tau_s = \frac{L_s}{R_s}$ and $\tau_r = \frac{L_r}{R_r}$ is the stator and rotor time constant respectively. The speed loop block diagram is illustrated in Figure 4 is based on the mechanical motor equation.

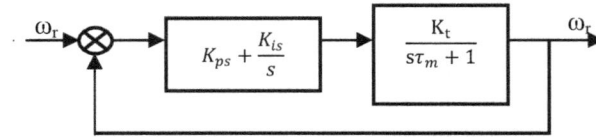

Figure 4. Simplified speed loop control

Where $\tau_m = \frac{J}{B}$ a is the motor mechanical time constant and torque constant, Kt is given as:

$$K_t = \frac{3}{2}\frac{P}{2}\frac{L_m^2}{L_s}\bar{I}_{sd} \tag{12}$$

The speed closed loop transfer function is given as below:

$$\frac{\omega_r^*}{\omega_r} = \frac{\left(K_{ps} + \frac{K_{is}}{s}\right)\left(\frac{K_t}{s\tau_m + 1}\right)}{1 + \left(K_{ps} + \frac{K_{is}}{s}\right)\left(\frac{K_t}{s\tau_m + 1}\right)} \tag{13}$$

The denominator of the general second order system is governed by;

$$s^2 + 2\varsigma\omega_n + \omega_n^2 \tag{14}$$

Where ω_n is the natural frequency of the closed-loop system and ς is the damping ratio. By comparing the denominator of the closed loop transfer function with Equation (14), the value of Kp and Ki can be determined. The gains of the PI controller are shown in Table 1. The values are obtain based on the equation above with ς is set at 1 and ω_n is set at 100Hz, 10Hz and 1Hz for torque loop, flux loop and speed loop respectively.

Table 1. PI Controller Parameters

PI Controller	Kp	Ki
Speed Controller	0.13	0.4252
Flux Controller	4.65	8.94
Torque Controller	13.4	197.45

3.2. PI Controller with Anti-Wind Up Scheme

The main objective of the AW scheme is to avoid the over value or saturation value in integrator which causes high overshoot and long settling time. Large step change or large external load disturbance applied causes the PI controller saturate. This windup phenomenon results in inconsistency between the real plant input and the controller output. In order to overcome the wind up problem, the integral state is separately controlled [4], [15], [18]. Thus, additional integral control in added to justified on the PI controller output is saturated or not based on the anti-windup structure in Figure 5.

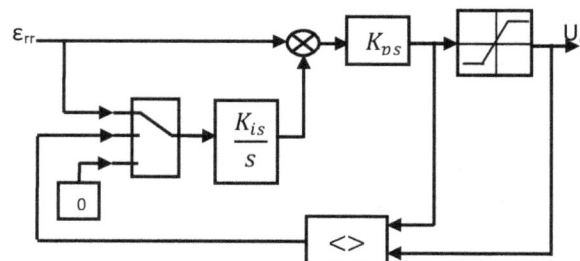

Figure 5. AWPI Conditional Integration Scheme

3.3. Anti Wind Up Scheme with Pre Filter

In order to have a pure second order system in the speed closed-loop, a pre-filter as shown in Equation (15) is added in series with the system [2], [3], [10].

$$G_{PF} = \frac{K_{is}}{k_{ps}S + K_{is}} \tag{15}$$

By inserting the pre-filter block, the behavior of the closed loop speed loop system is equal to the desired pure second order system. It is able to cancel the unwanted zero from the loop gain.

4.   RESULTS AND DISCUSSION

The performance comparison between PI controller, PI controller with anti-windup and pre-filter schemes is conducted using Dspace1103 controller. A three parallel insulated bipolar transistor (IGBT) intelligent power module (SEMiX252GB126HDs) are used for the inverter. The parameters of a 1.5kW induction motor are shown in Table 2. The voltage supply is set at rated voltage 380 Vrms and the switching frequency is set at 8kHz. The sampling time is 50μs. The tests are conducted to evaluate the performance of the motor under various speed operation demands and load disturbance rejection. Figure 6 shows the hardware experimental setup for TLI and FLI drive system.

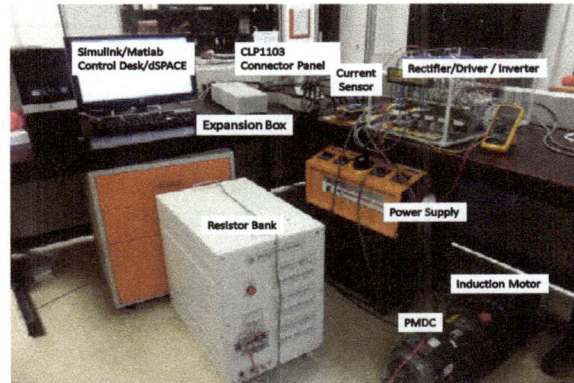

Figure 6. The hardware experimental setup

Table 2. Induction Motor Parameters

Motor Specifications	Value
Rated Voltage	380 V
Rated Frequency	50 Hz
Poles	4
Rated Speed	1430 rpm
Stator Resistance	3.45 Ω
Rotor Resistance	3.6141 Ω
Stator Inductance	0.3246 H
Rotor Inductance	0.3252 H
Magnetizing Inductance	0.3117 H
Inertia	0.02kgm^2
Viscous Friction	0.001 Nm/(rad/s)

4.1. Operation under Wide Speed Operation

This test is conducted during no load condition. For this experiment setup, 1.5kW Baldor three phase IM motor is coupled with 2.2kW Baldor permanent magnet DC machine. Incremental optical encoder is used to measure the shaft speed which has 500 pulses per revolution. For this test, the motor is required to operate at three different conditions which are standstill, forward direction and reverse direction at 500rpm, 1000rpm and 1400rpm operating speed. Every test is repeated for three times with different controllers' schemes. The first test is conducted using conventional proportional controller (PI) controller with limiters. Proportional controller with anti windup (AWPI) scheme for speed controller is applied for the second test condition. Then, the pre filter (PF) is added in cascade with the speed loop for the final experimental test.

Figure 7 shows the speed responses at 500rpm, 1000rpm and 1400rpm. The motor is required to operate from standstill to forward direction at 0.975s and reverse it direction at 4.226s. Based on the results, the motor tracks the command speed with almost zero speed error during steady state condition for all controllers. However, different transcient behaviours are notified such as rise time, percent overshoot and settling time. The details performance results from zero speed to forward direction are shown in Table 3 below:

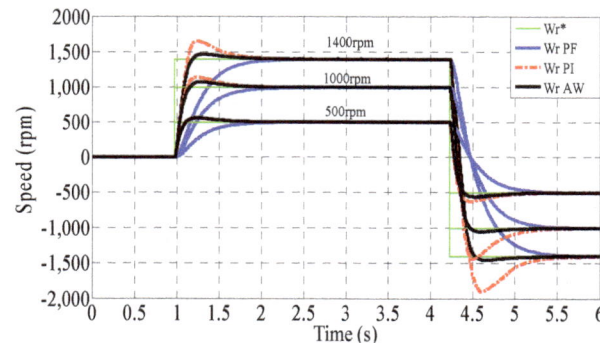

Figure 7. Speed response experiment results during standstill, forward and reverse direction at 500rpm, 1000rpm and 1400rpm speed operation (a) Overall performances (b) Closed up speed response at 500rpm (c) Closed up speed response at 1000rpm (d) Closed up speed response at 1400rpm

Table 3. Performance analysis of PF, PI and AW controller for forward Direction

Test Condition	Controller	%OS	Tr(s)	Ts(s)
500rpm	PI	11.6%	1.073	1.743
	AW	12.2%	1.077	1.718
	PF	0%	1.625	2.006
1000rpm	PI	14.3%	1.107	1.811
	AW	7.4%	1.091	1.727
	PF	0%	1.585	1.977
1400rpm	PI	18.3%	1.101	1.861
	AW	5.0%	1.130	1.640
	PF	0%	1.610	1.990

Meanwhile Table 4 shows the performance results from forward to reverse operation at 4.226s of the speed demand changed:

Table 4. Performance analysis of PF, PI and AW controller for reverse Direction

Test Condition	Controller	%OS	Tr(s)	Ts(s)
-500rpm	PI	24.2%	4.334	5.154
	AW	11.6%	4.362	5.022
	PF	0%	5.027	5.427
-1000rpm	PI	43.4%	4.383	5.333
	AW	5.3%	4.423	4.923
	PF	0%	5.041	5.411
-1400rpm	PI	35.64%	4.445	5.375
	AW	3.71%	4.447	4.897
	PF	0%	5.045	5.415

Based on the results, the AWPI and PI controller produced almost similar time rise response, Tr. Meanwhile, PF controller produces slower rise time response as well as the settling time. For the forward and reverse operations at 1400rpm demand, PI controller scheme recorded the highest percent overshoot with 18.3% and 35.64% respectively. No overshoot results from the pre filter controller for the demands changed. Meanwhile, the AWPI produce lower overshoot for those conditions at 5.0% and 3.71% respectively. The best characteristic of AWPI controller is it capabilities to produce lower percent overshoot while maintaining the rise time and improving the setting time. From the AWPI controller performance results, by increasing the speed range demand, the percent overshoot parameter is reduced compared to the conventional PI controller. In PI controller, a large step speed demand reference cause the output of the speed controller reaches the saturate limit of current, Iq. This anti-wind up phenomenon can be controlled by the AWPI scheme. The AWPI scheme control the integral parts from keep up integrating the error and the controller output from increased. The details analysis of the integral stage behaviors in PI and AWPI are explained based on the simulation results below.

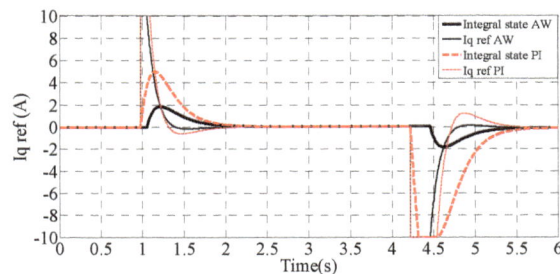

Figure 8. Simulation result comparing torque current response using PI and AWPI

Figure 8 shows the simulation results of current torque components response which compares between conventional PI and AWPI when a step function demand is applied from zero speed to 1400rpm in forward and reverse direction. Similar parameters, controllers and speed demand are used in this simulation and experiment. In the conventional PI scheme, the integral state becomes large at the start of linear region because it accumulates the speed error, even in saturation region. Thus, it produces excessive integral state results in a large overshoot. Meanwhile in the AWPI the integral state work only when the input and output saturation different is varnished. Therefore, it is able to reduce overshoot significantly and also maintains the rise time response, Tr results as PI controller. The speed control performance is much improved by AWPI scheme with regards to the large speed change.

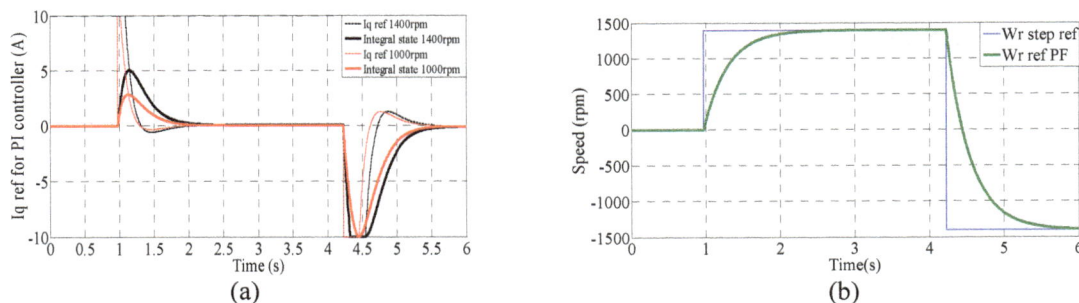

Figure 9. (a) Simulation result comparing torque current response using PI at 1000rpm and 1400rpm demand; (b) Reference step input demand and pre filter output demand

Figure 9(a) show the simulation results comparing torque current components for PI controller at 1000rpm and 1400rpm demand. During forward operation, the integral state output is 2.85A and 5A for 1000rpm and 1400rpm respectively. These results 14.3% and 18.3% of speed percent overshoot respectively. It means that, the higher changed of speed demand produce higher overshoot. However, during reverse operation, the percent of overshoot became 43.4% and 35.64% for 2000rpm and 2800rpm speed changed respectively. This result was affected by integral limiter integral limiter which is set at 10A. The integral state output is clamp at 10A for negative 1400rpm speed from 4.3s until 4.55s. This action, control the Iq reference demand from producing higher overshoot compared to negative 1000rpm demand change. As a result, lower percent overshoot with higher speed demand changes happened. The interesting part of the pre filter controller is it capabilities to maintain zero overshoot for all the speed demand range. Figure 9(b) shows the simulation comparison between step input reference and speed reference after the pre-filter process used as the reference signal in PF scheme. Due to the negative exponential speed reference demand by adding the PF, slower speed response results for PF scheme. This situation happened to all the step reference range demand and results no overshoot results but slower speed response.

Figure 10 show the torque current component, Iq and phase A current, Ia experiment results of the step response demands. From the stand still condition, the motor operated at 1400rpm in forward direction to 1400rpm reverse operation. Based on the result, almost similar torque current, Iq performance can be notify for PI and AWPI scheme. The torque current reach limited 10A set at the speed controller tremendously. PF scheme produce slower torque current response and only reach 4A amplitude during forward speed command. This also results in slower speed response of PF scheme. The similar results effect can be seen for the phase a stator current, Ia.

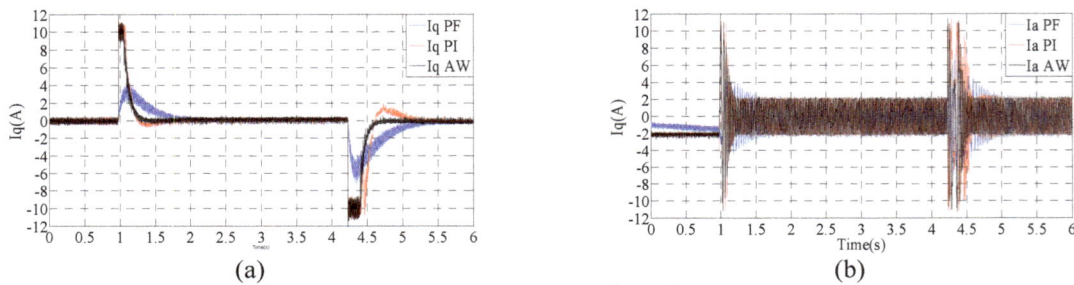

Figure 10. Experiment results during standstill, forward and reverse direction at 1400rpm speed operation, (a) Torque current response (b) *Phase A current response*

4.2. Operation under Load Condition

The load rejection capabilities of the design were investigated with the nominal load disturbance applied during rated speed operation as shown in Figure 11. The load disturbance operation is accomplished by using a DC machine attached with the load bank. The armature terminals of the permanent magnet DC machine are connected to the resistor bank. The external resistor of the DC machines is set to produce rated current load of IM. The motor is operated at 1400rpm and sudden rated load disturbance is applied at 2.5s. From the results, the speed dropped about 180rpm and recover from the undershoot situation within 1s for the entire schemes. It proven that, all the schemes has same capability of load disturbance rejection during rated load. However, the speed change is not large enough to turn-on the saturation condition of the anti-windup system.

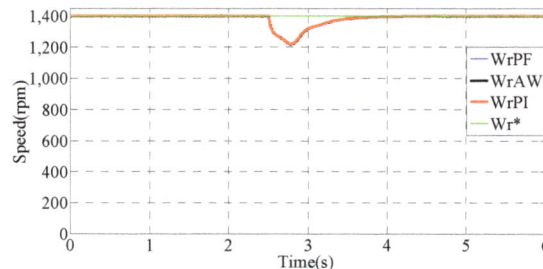

Figure 11. Experiment results with rated speed with nominal load dirstubunce

5. CONCLUSION

This paper presents the speed performance analysis of IFOC performance results between PI, AWPI and PF scheme control at low, medium and rated speed demand. All the three PI controller design are using second order system design approach. From the analysis, the AWPI is able to reduce overshoot problem by controlling the integral parts from keep up integrating the error at a set predetermine limiter. The result is more significant especially in term of percent overshoot reduction at the higher speed range demand. In addition, this great scheme is able to maintain the rise time and improving the setting time compared to the PI scheme. Meanwhile, PF scheme results slower speed and torque response performances. However, the PF is able to produce pure second order system in all speed range demands. It is able to get zero overshoot response with reasonable settling time response. Finally, for the load disturbance rejection ability, all schemes show similar performance capability to withstand the rated load disturbance.

ACKNOWLEDGEMENT

The authors would like to acknowledge their gratitude to Faculty of Electrical Engineering Universiti Teknikal Malaysia Melaka for providing the resources and support in this study.

REFERENCES

[1] GW Chang, G Espinosa-Perez, E Mendes, R Ortega, "Tuning rules for the PI gains of field-oriented controllers of induction motors," *IEEE Transactions on Industrial Electronics,*Vol. 47, No. 3, pp. 592-602, Jun. 2000.

[2] RC Dorf, RH Bishop, *Modern Control Systems, Tenth Edition,* Prentice Hall, 2005.

[3] J Espina, A Arias, J Balcells, C Ortega. Speed anti-windup PI strategies review for field oriented control of permanent magnet synchronous machines. *Compatibility and Power Electronics,* 2009; 279-285.

[4] C Jong-Woo, L Sang-Cheol. Antiwindup Strategy for PI-Type Speed Controller," *IEEE Transactions on Industrial Electronics.* 2009; 56(6): 2039-2046.

[5] Z Ibrahim, E Levi. A comparative analysis of fuzzy logic and PI speed control in high-performance AC drives using experimental approach. *IEEE Transactions on Industry Applications.* 2002; 38(5): 1210-1218.

[6] S Kuo-Kai, S Hsin-Jang. Variable structure current control for induction motor drives by space voltage vector PWM. *IEEE Transactions on Industrial Electronics.* 1995; 42(6): 572-578.

[7] K Satyanarayana, P Surekha, P Vijaya Prasuna. A new FOC approach of induction motor drive using DTC strategy for the minimization of CMV. *International Journal of Power Electronics and Drive Systems.* 2013; 3(2): 241-250.

[8] SN Nise. *Control System Engineering,* John Wiley & Sons, Inc., 2011.

[9] M Zelechowski, MP Kazmierkowski, F Blaabjerg. Controller design for direct torque controlled space vector modulated (DTC-SVM) induction motor drives. *IEEE International Symposium on Industrial Electronics,* 2005: 951-956.

[10] G Foo, CS Goon, MF Rahman. Analysis and design of the SVM direct torque and flux control scheme for IPM synchronous motors. *International Conference on Electrical Machines and Systems.* 2009: 1-6.

[11] KB Nordin, DW Novotny, DS Zinger. The Influence of Motor Parameter Deviations in Feedforward Field Orientation Drive Systems. *IEEE Transactions on Industry Applications.* 1985; IA-21(4): 1009-1015.

[12] C Bohn, DP Atherton. An analysis package comparing PID anti-windup strategies. *IEEE Control Systems.* 1995; 15(2): 34-40.

[13] S Hwi-Beon. New antiwindup PI controller for variable-speed motor drives.*IEEE Transactions on Industrial Electronics.* 1998; 45(3): 445-450.

[14] A Scottedward Hodel, CE Hall. Variable-structure PID control to prevent integrator windup. *IEEE Transactions on Industrial Electronics.* 2001; 48(2): 442-451.

[15] HB Shin. Comparison and evaluation of anti-windup PI controllers. *Journal of Power Electronics.* 2011; 11(1): 45-50.

[16] Z Ibrahim. Fuzzy Logic Control of PWM Inverter-Fed Sinusoidal Permanent Magnet Synchronous Motor Drives. *Ph. D. Dissertation, Liverpool John Moores.* 1999.

[17] S Eun-Chul, P Tae-Sik, O Won-Hyun, Y Ji-Yoon. A design method of PI controller for an induction motor with parameter variation. *Annual Conference of the IEEE Industrial Electronics Society.* 2003: 408-41.

[18] M Talib, Z Ibrahim, N Abdul Rahim, A Hasim. Implementation of Anti-Windup PI Speed Controller for Induction Motor Drive Using dSPACE and Matlab/Simulink Environment. *Australian Journal of Basic & Applied Sciences.* 2013; 7(1).

[19] ZM Salem Elbarbary, M Mohamed Elkholy. Performance analysis of field orientation of induction motor drive under open gate of IGBT fault. *International Journal of Power Electronics and Drive Systems.* 2013; 3(3): 304-310.

4

Lithium-ion Battery Charging System using Constant-Current Method with Fuzzy Logic based ATmega16

Rossi Passarella, Ahmad Fali Oklilas, Tarida Mathilda
Department of Computer Engineering, University of Sriwijaya, Palembang, Indonesia

ABSTRACT

Keyword:

Charging
Constant-current
Fuzzy
Lithium-Ion Battery

In this charging system, constant-current charging technique keeps the current flow into the battery on its maximum range of 2A. The use of fuzzy logic control of this charging system is to control the value of PWM. PWM is controlling the value of current flowing to the battery during the charging process. The current value into the battery depends on the value of battery voltage and also its temperature. The cutoff system will occur if the temperature of the battery reaches its maximum range.

Corresponding Author:

Rossi Passarella,
Department of Computer Engineering,
University of Sriwijaya,
Jln. Palembang-Prabumulih, km 32. Inderalaya, OganIlir, Sumatera-selatan, Indonesia 30662
Email: passarella.rossi@gmail.com

1. INTRODUCTION

There are many charging methods for lithium battery such as constant-current method, constant-voltage method, conventional five-stage or proposed fuzzy-based algorithm method [1]. Charging Lithium battery with constant-current method is a technique to keep the value of the current when it flows into battery, while the value of its voltage is charging [2]-[5]. Even though the value of the current is fluctuating, but in this charging system, the maximum value will be 2Ampere. Here, the changing value of battery voltage is from range 2.7 volt to 4.2 volt.

The addition of fuzzy logic control of this Lithium battery charging system is the control of current flow into the lithium battery, so that it will meet its input and output requirements.

There are two inputs in this charging system, which are temperature and voltage of lithium battery. Temperature is the most vital parameter in lithium battery security that affected battery's health. The lithium battery is easy to explode when it is overcharging that caused by over temperature.

The objective of this study is the current flows into the Lithium battery can be controlled, by changing the temperature and increasing the voltage of the battery.

2. CHARGING METHOD

There are many kinds of charging methods for battery, example constant–current, constant-voltage, and five-stage Li-ion battery charger [1]-[5]. During the constant current phase, the primary task of battery management is to control the flow of current to the maximum permissible battery current [4]-[5].

Battery use in this charging system is Panasonic CGR18650CG [6]. The specification of the battery shown in Table 1.

Table 1. Battery Specification

Measurements		Quantity
Nominal Voltage		3.6 v
Nominal Capacity	Minimum	2.150 mAh
	Typical	2.250 mAh
Dimension	Diameter	18.6 mm
	Height	65.2 mm

This charging system is using MOSFET's transistor as an active instrument. MOSFET is an instrument which read the electric signal and controls the output voltage from the charger system onto the battery.In this charging system, MOSFET is use because it has better durable than other common transistors. This MOSFET can resist the flow of the current up to 10Ampere. In charging system the PWM (Pulse Width Modulation) technology is applied to set the function of charging system to battery.

3. RESULTS AND ANALYSIS

The key design of software from this charging system is Fuzzy Algorithm. The Fuzzy inference system of this charging system is Sugeno's model. On Sugeno's model, to bring out the output we need four steps, which: forming of Fuzzy's set (fuzzification), function of implication, evaluation of rules, and defuzzification [7]. The evaluation rules use Max-Min mechanism and the defuzzification step use Center of Average (CoA) method.The flowchart of charging system is shown in Figure 1.

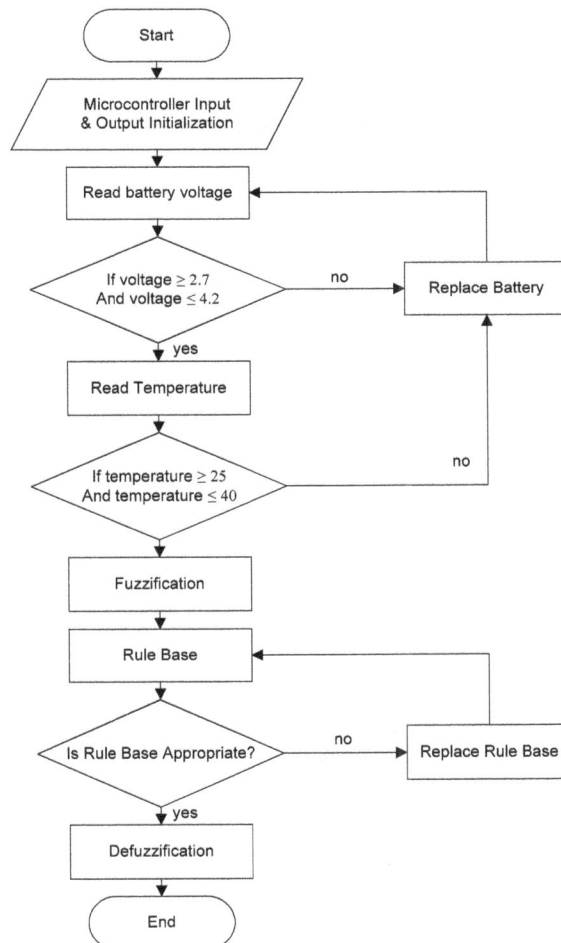

Figure 1. Flowchart Charging System

3.1. Fuzzification

This system uses two inputs which are voltage and temperature of the battery. First, ADC microcontroller read battery's voltage with sensor and set the linguistic form. Linguistic forms of battery voltage and battery temperature shown in Table 2 and Table 3.

Table 2. Input voltage

VOLTAGE (V)	LINGUISTIC
2.7 – 3.2	Low2
3.0 – 3.6	Low1
3.2 – 3.8	Normal
3.6 – 4.0	High1
3.8 – 4.2	High2

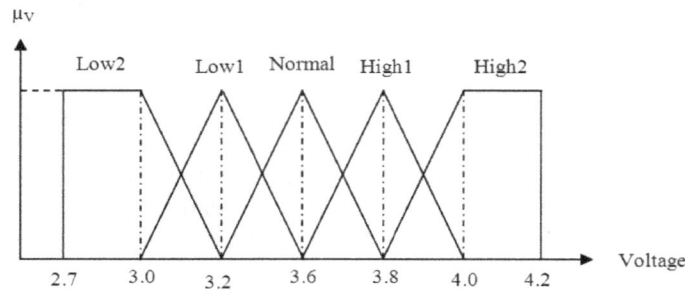

Figure 2. Fuzzy Voltage's range

Figure 2 shows the sets of voltage's range. It consist of five areas, starting from 2,7 volt until 4,2 volt, naming low2, low1, normal, high1, high2. System will run cut-off, once the voltage of the battery reach above 4,2 volt.

Table 3. Input - temperature

Temperature (^{0}C)	Variabel Linguistik
18-24	Inc1
21-29	Inc2
24-34	Inc3
29-37	Inc4
34-40	Inc5

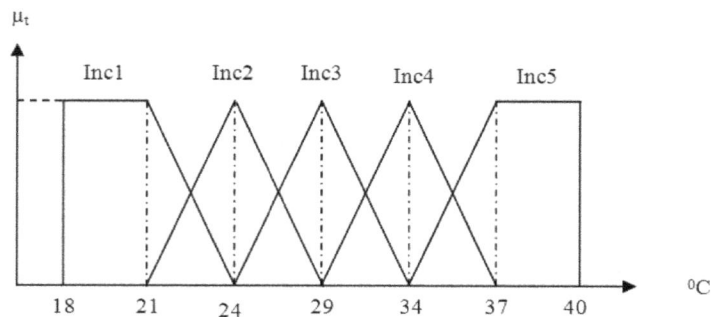

Figure 3. Fuzzy Temperature's range

Figure 3 shows the sets of temperature's range. It also consists of five areas that starting at 18°C-40°C. System will also run cut-off, once the temperature reach above 40°C.

3.2. Rule Base

Rule base of this system is from each input of fuzzy logic. So that, there will be 25 rules, and shown in Table 4.

Table 4. Rule base

No	Input		Output
	Voltage	Temperature	Current
1	Low2	Inc1	Rapid
2	Low2	Inc2	Rapid
3	Low2	Inc3	Rapid
4	Low2	Inc4	Rapid
5	Low2	Inc5	Normal
6	Low1	Inc1	Rapid
7	Low1	Inc2	Rapid
8	Low1	Inc3	Rapid
9	Low1	Inc4	Rapid
10	Low1	Inc5	Normal
11	Normal	Inc1	Rapid
12	Normal	Inc2	Rapid
13	Normal	Inc3	Rapid
14	Normal	Inc4	Normal
15	Normal	Inc5	Normal
16	High1	Inc1	Normal
17	High1	Inc2	Normal
18	High1	Inc3	Normal
19	High1	Inc4	Slow
20	High1	Inc5	Slow
21	High2	Inc1	Slow
22	High2	Inc2	Slow
23	High2	Inc3	Slow
24	High2	Inc4	Slow
25	High2	Inc5	Slow

3.3. Mechanism of Inference

Mechanism of inference in this system transform into three ranges of percents, in duty cycles of PWM will run, which are rapid, normal and slow and shown in Table 5, and its formula in Equation (1)

$$\%PWM = \frac{\sum_{1=0}^{n} v_i \mu_v(v_i)}{\sum_{i=0}^{n} \mu_v(v_i)} \qquad (1)$$

Where : % PWM is output, v_i is *crisp's value* of i's element, $\mu_v(v_i)$ is degree of every elements in Fuzzy's set of V. V isuniverse of Fuzzy, and n is quantization.

Table 5. Mechanism Inference

Duty Cycle (%)	Linguistic	Information
30	Rapid	Max1
60	Normal	Max2
90	Slow	Max3

3.4. Defuzzification

Defuzzification of this system is using CoA (center of Average), by formula in Equation (2):

$$y = \frac{\sum y \mu_R(y)}{\sum \mu_R(y)} \qquad (2)$$

Where : y is crisp's value and μ_R (y) is membership of y.

3.5. Pulse Width Modulation (PWM)

Pulse Width Modulation (PWM) is a method for using pulse width to encode or modulate a signal. The width of each pulse is a function of the amplitude of the signal. While ADC detect the battery voltage and LM35 detect the changing of temperature, microcontroller will deliver and group those inputs into Fuzzy's set. Furthermore, microcontroller will control the IC to deliver the PWM signal into MOSFET series. The value of the current will depend on the mathematics calculation in the microcontroller.

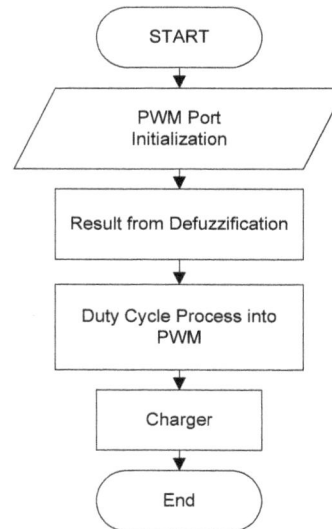

```
        START
          │
   PWM Port
   Initialization
          │
   Result from Defuzzification
          │
   Duty Cycle Process into
   PWM
          │
       Charger
          │
        End
```

Figure 6 .Flowchart of PWM

4.   RESULTS
4.1. First Experiment

In the first experiment (Figure 7), the room temperature was set at 25°C, in 2 hours (7200s) and the initial battery voltage was 2.7 volts. In the 1s , the battery temperature was 25.1°C, the current inflows was recorded at 2 amperes. At 238 second the voltage increase up to 2.8 volts, and the temperature was recorded at 25.5° C with a flow to the battery at 1.9 amperes. The decrease in flow occurs due to the temperature rise. At 469 second, the voltage increase up to 2.9 volts with battery temperature was 26.9° C and current was at 2 Amperes. At 3.0 volts, temperature was 27.3°C and the current was 1.9 Amperes. At 3.1 volts voltage of battery on 991 second, the temperature was at 29.2° C with current flows into the battery at 2 Amperes. At 1135 second voltage rise to 3.2volts with a recorded temperature of 29.8° C and the current flow of 1.8 Ampere.

It can be concluded that Fuzzy logic work when temperature is rising in the battery current flows. When the current flow increases, the temperature will increase, so the next current flow can be reduced, and the temperature can be decreased.

4.2. Second Experiment

In the second experiment (Figure 8), the room temperature was 25°C, experiments approximatelywith in 2 hours (7200s) with initial battery voltage at 2.7 volts. In the 1s, the temperature was 26°C, the flows of current was 2 amperes. At 240 second the voltage increased up to 2.8 volts, and the temperature was at 26.1°C with current flow to battery was 2 Ampere. At 500 second, the voltage increased up to 2.9 volts and the temperature was 26.3°C with current at 2 ampere. At 3.0 volts, temperature was 26.5°C and the current flows at 2 amperes. At 3.1 volts at 870 second, the temperature was at 27.1°C with current flows of 1.9 ampere. At 1019 second, voltage up to 3.2 volt and temperature was 27.2°C with the current flows at 1.9 amperes.

Similar to the first experiment: in conclusion the fuzzy logic control works similar to the first experiment.

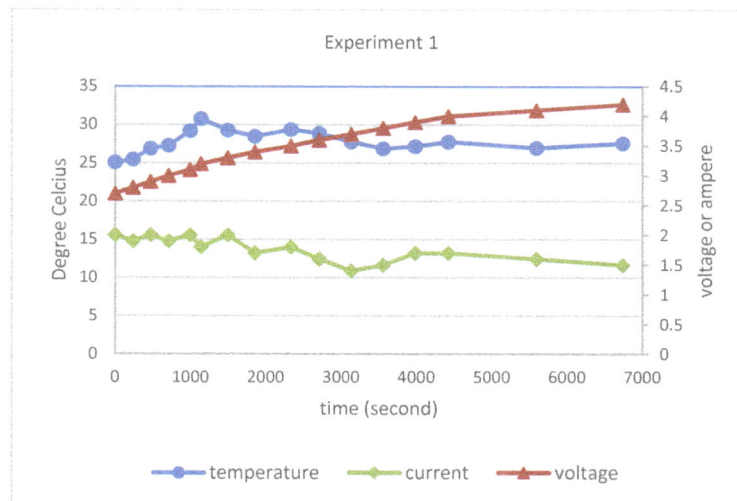

Figure 7. Graph of First Experiment result

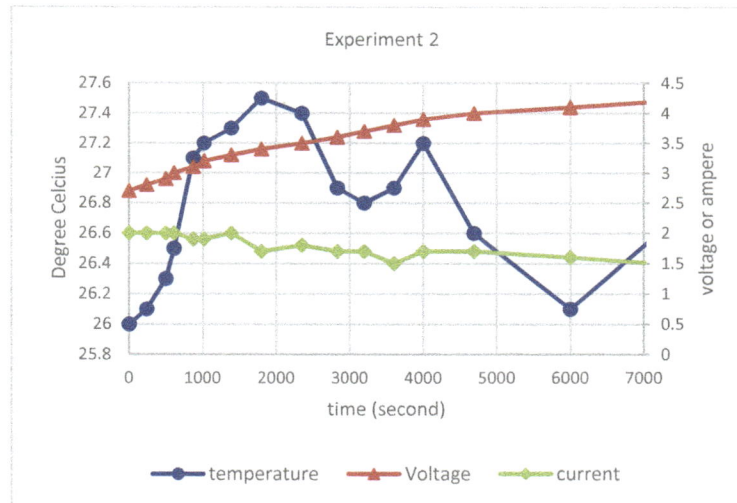

Figure 8. Graph of Second Experiment result

4.3. Third Experiment

In the third experiment (Figure 9), the room temperature was 25°C, approximatelywith in 2 hours (7200s) the initial battery voltage at 2.7 volts. In the first second, the temperature was at 25.1°C, the current flows to battery was 2 amperes. At 294 second the voltage increased up to 2.8 volts, and the temperature at 25.2°C with current flowed to the battery at 1.9 amperes . At 504 second the voltage increased to 2.9 volts with temperature was 25.4°C and the current was 2 amperes. At 3.0 volts , temperature 25.3°C and the current flow at 2 amperes. At 3.1 volts at 890 second, the temperature was 25.6°C and current flow at 2 Ampere. In 121 second the voltage rise to 3.2 volts and temperature was 26.1°C with the current flows at 1.8 Amperes.

In the charging system of the lithium ion battery, the critical parameter that should be considered is temperature, due to this the comparison between experiments was plot in the graph as shown in Figure 10. The results show that the temperature batteries are below the data sheet.

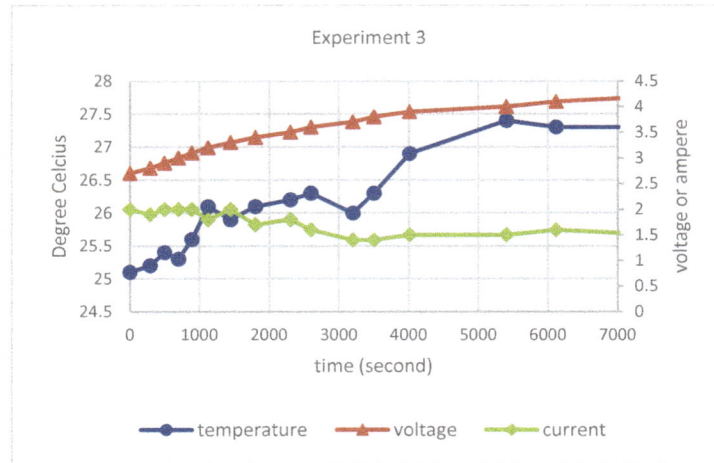

Figure 9. Graph of Third Experiment result

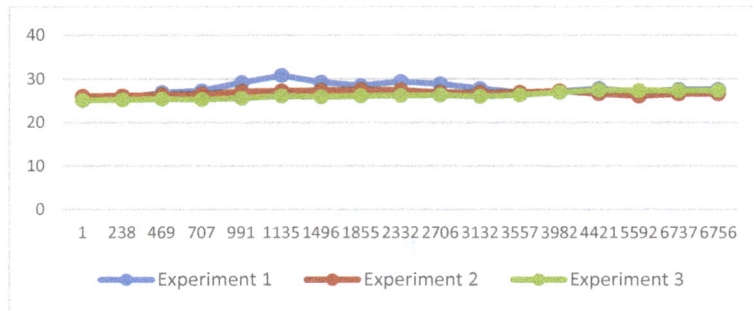

Figure 10. Comparison of the temperature from 3 experiments

At the end of the experiment, the averages of each battery parameters are shown in Table 6.

Table 6. The Average Battery parameter in Experiment

Parameter	BATTERY	
	Temperature (°C)	Current (A)
Experiment 1	27.75	1.65
Experiment 2	26.77	1.68
Experiment 3	26.21	1.62
Average	26.91	1.65

From the Table 6 shows that the average value of experiments for each battery starting from the first to the third trial are: temperatures of 26.91°C and current of 1.65 Amperes. In this experiment the current flows into the Lithium battery can be controlled, by changing the temperature and increasing the voltage of the battery.

5.   CONCLUSION

a)   This system consists of two parts, which are: microcontroller series function to calculate the Fuzzy, and MOSFET function for charger series.

b) The design algorithm to control the flows of current use PWM. The output from the microcontroller calculation will be delivered through MOSFET to control the value of the current flowing into the Lithium battery.

c) The total of current flows into Lithium battery is affected by the value of voltage and temperature while it is charging.

d) Fuzzy is still working despite the temperature of the Lithium battery changing. The voltage of the battery will constantly rises until it reaches 4.2 volt.

e) The value of the current flowing into the lithium battery is depending on the value of the temperature of the battery, as it is formulated in the rule base of Fuzzy.

The average temperature of the lithium battery while charging process running is 26°C and the average of the current flowing into the battery is 1,75A.

ACKNOWLEDGEMENTS

This work was supported by Department of Computer Engineering, Faculty of Computer Science. University of Sriwijaya.

REFERENCES

[1] HoushyarAsadi, et al. Fuzzy Logic Control Technique in Li-Ion Battery Charger. International conference on electrical, electronics and civil engineering (iceece'2011) pattayadec. 2011; 179-183.

[2] Huang, Jia-Wei, et al. Fuzzy-control-based five-step Li-ion battery charger. In Power Electronics and Drive Systems. PEDS. International Conference on. IEEE, 2009; 1547-1551

[3] Asadi, Houshyar. et al. Fuzzy-control-based five-step Li-ion battery charger by using AC impedance technique." In Fourth International Conference on Machine Vision (ICMV 11), pp. 834939-834939. International Society for Optics and Photonics. 2012.

[4] Hsieh, Ching-Hsing. Research on the Five Step Charging Technique for Li-ion Batteries Using Taguchi Method and Fuzzy Control. PhD diss., 2011.

[5] Manoj, Niranjan Kumar, Vijay Pal Singh. Fuzzy Logic Based Battery Charger for Inverter." International Journal of Engineering. 2013; 2(7).

[6] Cgr 18650-cg. Datasheet lithium-ion rechargeable cell. Panasonic corporation energy company. Februari. 2010. Available:www.industrial.panasonic.com/wwwdata/pdf2/aca4000/aca4000ce234.pdf&sa=u&ei=gq9vuv6hbszfkwf2 yhobq&ved =0cbmqfjad&usg=afqjcnemwazpsmqr9zuhrkvgmha2367jda.

[7] Passarella, Rossi, et al. PerancanganSistemPenjadwalanBateraiBerbasisLogika Fuzzy MenggunakanMikrokontroler ATMega16. KonferensiNasionalInformatika (KNIF). 2013: 54-58.

APPENDIX

Experiment #1

Time (s)	Battery		
	Time (°C)	Volt (V)	Current (A)
1	25.1	2.7	2
238	25.5	2.8	1.9
469	26.9	2.9	2
707	27.3	3	1.9
991	29.2	3.1	2
1135	30.8	3.2	1.8
1496	29.3	3.3	2
1855	28.5	3.4	1.7
2332	29.4	3.5	1.8
2706	28.9	3.6	1.6
3132	27.8	3.7	1.4
3557	26.9	3.8	1.5
3982	27.2	3.9	1.7
4421	27.8	4.0	1.7
5592	27	4.1	1.6
6737	27.6	4.2	1.5
6756	27.6	4.2	0.0	Cut-off
Mean	27.75		1.65

Experiment #2

Time (s)	Battery		
	Temp (°C)	Volt (V)	Current (A)
1	26	2.7	2
240	26.1	2.8	2
500	26.3	2.9	2
610	26.5	3.0	2
870	27.1	3.1	1.9
1019	27.2	3.2	1.9
1393	27.3	3.3	2
1802	27.5	3.4	1.7
2350	27.4	3.5	1.8
2830	26.9	3.6	1.7
3201	26.8	3.7	1.7
3605	26.9	3.8	1.5
4002	27.2	3.9	1.7
4690	26.6	4.0	1.7
6001	26.1	4.1	1.6
7201	26.6	4.2	1.5
7220	26.6	4.2	0.0	Cut-off
Mean	26.77	-	1.68

Experiment #3

Time (s)	Battery		
	Temp (°C)	Volt (V)	Current (A)
1	25.1	2.7	2
294	25.2	2.8	1.9
500	25.4	2.9	2
699	25.3	3.0	2
890	25.6	3.1	2
1121	26.1	3.2	1.8
1444	25.9	3.3	2
1801	26.1	3.4	1.7
2305	26.2	3.5	1.8
2599	26.3	3.6	1.6
3200	26	3.7	1.4
3501	26.3	3.8	1.4
4013	26.9	3.9	1.5
5404	27.4	4.0	1.5
6120	27.3	4.1	1.6
7580	27.3	4.2	1.5
	27.3	4.2	0.0	Cut-off
Mean	26.15	-	1.73

5

Enhancement of Power Quality by an Application FACTS Devices

Prashant Kumar

Departement of Electrical Engineering, Ashokrao Mane Group of Institutes, Shivaji University, Maharshtra, India

ABSTRACT

The paper narrates widespread use of electrical energy by modern civilization has necessitated producing bulk electrical energy economically and efficiently. The Flexible AC Transmission system (FACTS) is a new technology based on power electronics, which offers an opportunity to enhance controllability, stability, and power transfer capability of AC transmission systems. Here SVC has been developed with the combination of TCSC and TCR. The paper contains simulation models of Thyristor controlled Series Capacitor (TCSC) and Thyristor controlled Reactor (TCR)-based Static VAR Compensator (SVC) which are the series and shunt Flexible AC Transmission Systems (FACTS) devices. The fact devices are designed by considering the line losses and their stability. The design and simulations of TCSC and TCR-based SVC shows the effectiveness of result using the MATLAB/Simulink. The designed system will try to reduce the voltage drops and electrical losses in the network without the possibility of transient especially in case of long transmission system. Student feedback indicates that this package is user-friendly and considerably effective for students and researchers to study theory of controlled reactor compensators, series capacitor compensator, and the reactive power control and voltage regulation

Keyword:

Facts Controller
Matlab/Simulink
Thyristor Controlled Reactor
Thyristor Controlled Series
Capacitor

Corresponding Author:

Prashant Kumar,
Departement of Electrical Engineering,
Ashokrao Mane Group of Institutes, Shivaji University,
Kolhapur, Shivaji University, Maharshtra, India.
Email: prashant2685@gmail.com

1. INTRODUCTION

Since from the last decade we facing problems to meet the demand of energy as because of industrial growth of a nation requires increased consumption of energy, particularly electrical energy. This has led to increase the generation and transmission facility to meet the increasing demand. For generation, transmission, distribution and utilization of electrical energy, 3 phase AC systems are used universally. It is beneficial to use AC system because of its features like reduction of electrical losses, increasing transmission efficiency and capacity, better voltage regulation, reduction in conducting material, flexibility for growth and possibility of interconnection. FACTS Controller is defined as a power electronic-based system and the other static equipment that provide control of one or more AC transmission system parameters. This paper describes basic types of FACTS controllers. Use of HVAC is economical till breakeven point having distance around 800km only, after this point HVAC become much costlier. The corona effects tend to be highly significant for HVAC and the design of AC conductors based on the corona limitations gives a cross-section much larger than that with respect to economical power transfer limits. Also it needs heavy supportive structures that lead to erection difficulties. Stability of AC networks is very low due to the line inductive reactance. Voltage control is difficult for long line due to series inductance and capacitance and requires more complex circuits. Hence the reliability issue needs to be addressed seriously. To compensate drawback of HVAC mention above we require HVDC transmission system. The reliability of HVDC is quite good. A DC line can carry as much power with two conductors as an AC line with three conductors of the same line. In HVDC there are less power losses, absence of skin effect, less corona effect as compared to AC. No compensation is required, no limits for power transfer and control of voltage is easier.It is not possible and economical to replace already exists AC transmission system by HVDC up to certain breakeven level of power as well as distance. The converter required at both the end of the line have proved to be reliable but they are much more expensive than the conventional equipments. HVDC converters need complex cooling systems. Maintenance of insulation is more in HVDC. The DC system cannot be employed for a distribution, sub transmission and the backbone transmission. Voltage transformation is not easier in case of DC; hence it has to be accomplished on the AC side of system. So it is not much suitable for transmission interconnections.

Power system engineers are currently facing challenges to increase the power transfer capabilities of existing transmission system. This is where the Flexible AC Transmission Systems (FACTS) technology comes into effect. With relatively low investment, compared to new transmission or generation facilities, the FACTS technology allows the industries to better utilize the existing transmission and generation reserves, while enhancing the power system performance. Moreover, the current trend of deregulated electricity market also favors the FACTS controllers in many ways. FACTS controllers in the deregulated electricity market allow the system to be used in more flexible way with increase in various stability margins. FACTS controllers are products of FACTS technology; a group of power electronics controllers expected to revolutionize the power transmission and distribution system in many ways. The FACTS controllers clearly enhance power system performance, improve quality of supply and also provide an optimal utilization of the existing resources. Thyristor Controlled Series Compensator (TCSC) is a key FACTS controller and is widely recognized as an effective and economical means to enhance power system stability. In this paper an overview to the general types of FACTS controllers is given along with the simulation of TCSC FACTS controller using SIMULINK. Analysis of the simulated TCSC shows similar functions as a physical one.

2. THEORY OF THYRISTOR-CONTROLLED AND THYRISTOR-SWITCHED REACTOR (TCR AND TSR)

An elementary single phase Thyristor-controlled reactor (TCR) is shown in Figure 1.

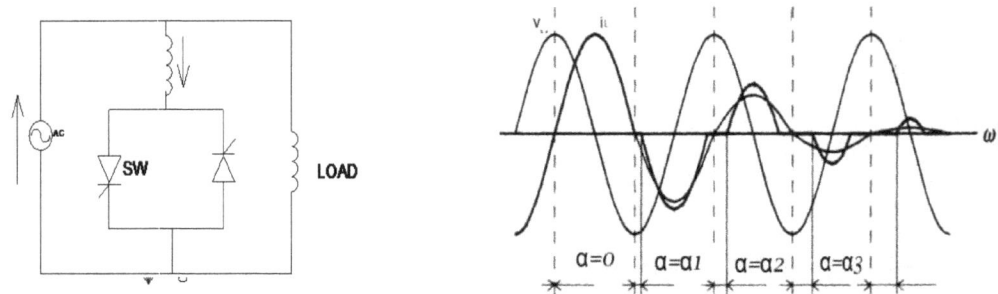

Figure 1. Basic Thyristor-controlled reactor (TCR) firing Delay Angle Control and operating waveform

It consists of a fixed reactor of inductance L, and a bidirectional thyristors valve or a switch sw. The current in the reactor can be controlled from maximum to zero by the method of firing delay angle control. That is, closure of the thyristors valve is delayed with respect to the peak of the applied voltage in each half cycle and thus the duration of the current conduction intervals is controlled. This methods of current control is illustrated separately for the positive and negative half cycles in Figure 1, where the applied voltage v and the rector current $i_L(\alpha)$, at zero delay angle and at arbitrary α delay angle, are shown.

The current in the reactor can be expressed with, $v(t) = V \cos \omega t$ as follow:

$$i_L(t) = \frac{1}{L} \int_\alpha^{\omega t} v(t) dt = \frac{V}{\omega L} (\sin \omega t - \sin \alpha) \tag{1}$$

Where V is the amplitude of the applied ac voltage, L is the inductance of the thyristor-controlled reactor, and ω is the angular frequency of the applied voltage.The TCR can control the fundamental current continuously from zero (valve open) to the maximum (valve closed).

SVC Configurations: Providing reactive shunt compensation with shunt-connected capacitors and reactors is a well-established technique to get a better voltage profile in a power system [2]. The basic form of reactive power compensation required, to compensate reactive power loads, is the fixed shunt capacitors being well distributed across the network and located preferably closed to the loads. This would ensure reasonable voltage profile during steady state condition. However, this may not be adequate to ensure stability under overload or contingency conditions. Shunt capacitors are inexpensive but lack dynamic capabilities, thus some form of dynamically controlled reactive power compensation becomes essential. The phase angle between the end voltages, determined by the real component of the line current, is not affected by the shunt compensation. Similarly, adding a reactor instead of a capacitor in shunt will reduce the voltage. Instead of mechanical switching (using circuit breakers) of these devices, we can use thyristor valves, thereby increasing the control capability radically. This approach is called static VA R compensation (SVC).

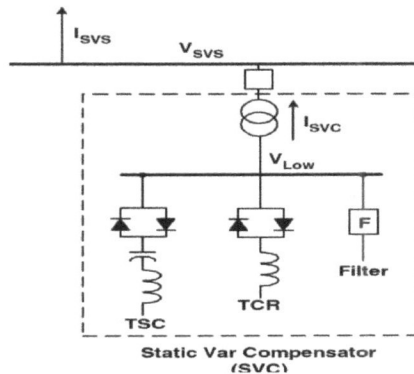

Figure 2. Basic configuration static var compensator

SVC can be of one of the following types:
1. Thyristor controlled Reactor (TCR)
2. TCR plus Fixed Capacitor (FC)
3. Thyristor switched Capacitor (TSC)
4. TSC plus TCR

Figure 2 is a one-line diagram of a typical static VAR system for the transmission application. TSC plus TCR is very popular and most effective. Fig 3 gives the general idea of realization of SVC using TSC plus TCR scheme. The idea is to sense the voltage of the line and keep it stable by introducing capacitance or inductance in the circuit.

The Figure 3 presents an equivalent circuit of the TCR. The TCR consists of two thyristor in anti-parallel, a reactor. Also in the three phase applications, the basic TCR elements are connected in delta. A SMIB system with a TCR based SVC as shown in Figure 4. The shunt controller is injects current into the line at point of common coupling (PCC). The main function of TCR is to current controlled by controlling the firing angles of thyristor. So obviously power can be controlled. Since control can be achieved in every cycle of the voltage waveform by (controlling the conduction time of thyristors), the control is very fast and accurate.

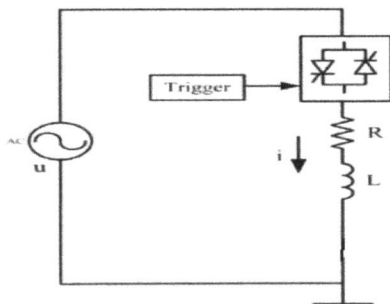

Figure 3. Basic structure of TCR

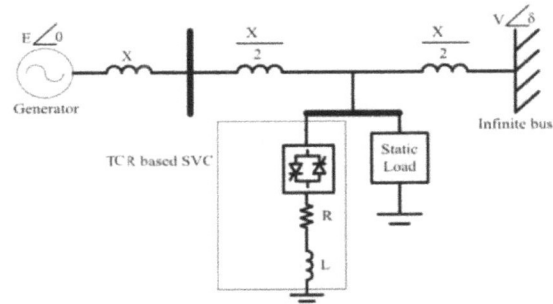

Figure 4. SMIB system with a TCR based SVC

3. THEORY OF THYRISTOR CONTROLLED SERIES CAPACITOR (TCSC)

The basic conceptual TCSC module comprises a series capacitor, C, in parallel with a thyristor-controlled reactor, LS, as shown in Figure 5(a). A TCSC is a series-controlled capacitive reactance that can provide continuous control of power on the ac line over a wide range. From the system viewpoint, the principle of variable-series compensation is simply to increase the fundamental-frequency voltage across fixed capacitor (FC) in a series- compensated line through appropriate variation of the firing angle [4]. This enhanced voltage changes the effective value of the series-capacitive reactance.

Simple understanding of TCSC functioning can be obtained by analysing the behaviour of a variable inductor connected in parallel with an FC, as shown in Figure 5(b).

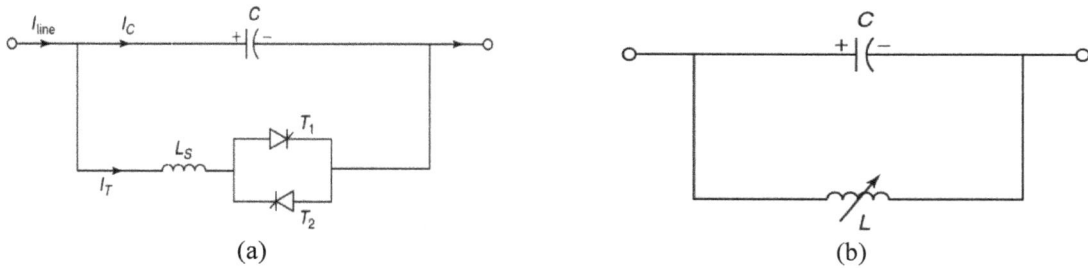

(a)   (b)

Figure 5. TCSC Module (a) Basic module and (b) A Variable inductor connected in shunt with FC

The equivalent impedance, Zeq, of this LC combination is expressed as:

$$Z_{eq} = (j\frac{1}{wc}) \| (jwL)$$

(2)

The impedance of the FC alone, however, is given by $-j(1 / \omega C)$.

If $\omega C - (1 / \omega L) > 0$ or, in other words, $\omega L > (1 / \omega C)$, the reactance of the FC is less than that of the parallel-connected variable reactor and that this combination provides a variable-capacitive reactance are both implied. Moreover, this inductor increases the equivalent capacitive reactance of the LC combination above that of the FC. If $\omega C - (1 / \omega L) = 0$, a resonance develops that results in an infinite capacitive impedance an obviously unacceptable condition. If, however, $\omega C - (1 / \omega L) < 0$, the LC combination provides inductance above the value of the fixed inductor. This situation corresponds to the inductive mode of the TCSC operation. In the variable-capacitance mode of the TCSC, as the inductive reactance of the variable inductor is increased, the equivalent-capacitive reactance is gradually decreased. The minimum equivalent-capacitive reactance is obtained for extremely large inductive reactance or when the variable inductor is open-circuited, in which the value is equal to the reactance of the FC itself. The behaviour of the TCSC is similar to that of the parallel LC combination. The difference is that the LC-combination analysis is based on the presence of pure sinusoidal voltage and current in the circuit, whereas in the TCSC, because of the voltage and current in the FC and thyristor-controlled reactor (TCR) are not sinusoidal because of thyristor switching.

The series compensation provided by the TCSC can be adjusted rapidly to ensure specified magnitudes of power flow along designated transmission lines. This condition is evident from the TCSC's efficiency, that is, ability to change its power flow as a function of its capacitive-reactance setting:

$$P_{12} = \frac{V_1 V_2}{(XL - XC)} \sin \partial$$

(3)

Where:
P12= the power flow from bus 1 to bus 2
V1, V2 = the voltage magnitudes of buses 1 and 2, respectively
XL = the line-inductive reactance
 XC = the controlled TCSC reactance combined with fixed-series- capacitor reactance
d = the difference in the voltage angles of buses 1 and 2

This change in transmitted power is further accomplished with minimal influence on the voltage of interconnecting buses, as it introduces voltage in quadrature. In contrast, the SVC improves power transfer by substantially modifying the interconnecting bus voltage, which may change the power into any connected passive loads. The freedom to locate a TCSC almost anywhere in a line is a significant advantage. Power-flow control does not necessitate the high-speed operation of power-flow control devices. Hence discrete control through a TSSC may also be adequate in certain situations. However, the TCSC cannot reverse the power flow in a line, unlike HVDC controllers and phase shifters.

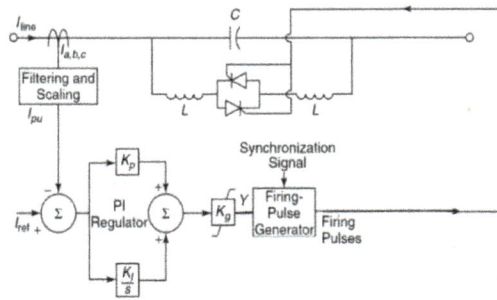

Figure 5. TCSC controller model

4. RESULTS AND ANALYSIS

4.1. Simulation Model of Shunt Connected TCR on 1-phase Line

In first case study, a single-phase system with TCR is considered. The single-phase transmission line is simulated using MATLAB/Simulink. In the simulation study, it is assumed that the current source peak amplitude = 50A, It is connected to capacitive load through Pi transmission line, and the TCR controller is shunt connected to the transmission line. Analysis and comparison are done based on the results obtained from the line power of the single-phase line employing the shunt controller, in terms of output power waveforms, output current waveform, output voltage waveforms. The Design of TCR is as shown in Figure6. The PWM technique is used to control the firing pulses to gates of both thyristors.

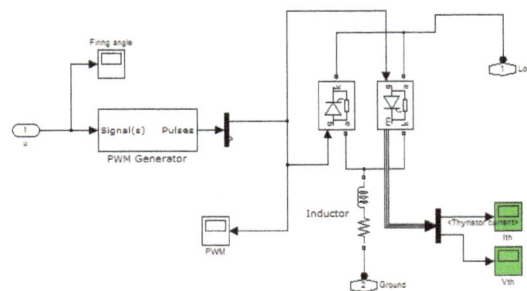

Figure 6. Design Model of TCR

4.2. Simulation Model of Series Connected TCSC on 1-phase Line

In second case study, a single-phase system with TCSC is considered. The single-phase transmission line is simulated using MATLAB/Simulink. In the simulation study, it is assumed that the current source peak amplitude = 50A, It is connected to inductive load through Pi transmission line, and the TCSC controller is series connected to the transmission line. Analysis and comparison are done based on the results obtained from the line power of the single-phase line employing the series controller, in terms of output power waveforms, output current waveform, output voltage waveforms. The Design of TCSC is as shown in Figure 7. The PWM technique is used to control the firing pulses to gates of both thyristors.

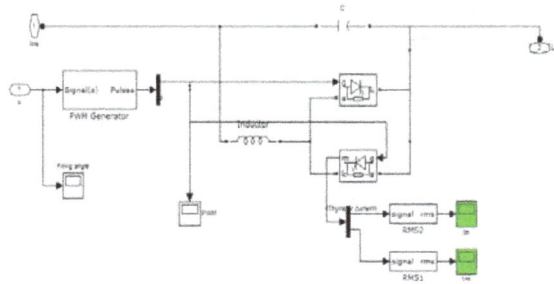

Figure 7. Design Model of TCSC

4.3. Simulation Results of Shunt Connected TCR

In single phase transmission line, Input power gets 100KW by simulating model of TCR as shown in Figure 8. Then power decreases in transmission line due to present line parameters. But using TCR shunt controller, the power transfer capacity of line increases and at the load side power gets 50KW as shown in Figure 9. Also by using shunt controller voltage control is possible. The analysis shows that if compensation is provided through TCR controller then the system attains stability at a faster rate.

Figure 8. Input Power in RMS

Figure 9. Load Power in RMS

4.4. Simulation Results of Series Connected TCSC

In single phase transmission line, Input power gets 4MW by simulating model of TCSC as shown in Figure 10. Then power decreases in transmission line due to present line parameters. But using TCSC series controller, the power transfer capacity of line increases and at the load side power gets 3.9MW as shown in Figure 11. The analysis shows that if compensation is provided through TCSC controller then the system attains stability at a faster rate.

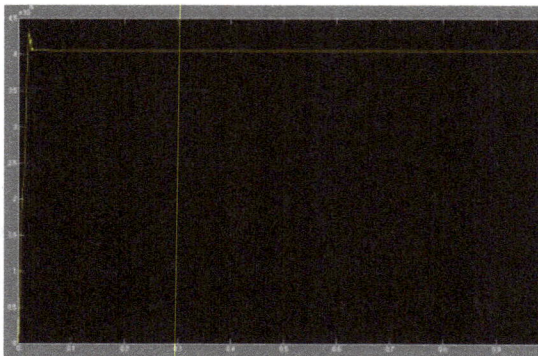

Figure 10. Input Power in RMS

Figure 11. Load Power in RMS

4.5. PID Controller is Designed for TCR and TCSC

The PID controller is a very simple controller, but the major drawback is that there is no analytical way of finding the optimal set of parameters K_P, K_I, and K_D. The conventional Proportional Integration Derivative (PID) structure remains the controllers of choice in many industrial applications because of its structural simplicity, reliability and the favourable ratio between performance and cost. Beyond these benefits, this controller also offers simplified dynamic modelling, lower user skill requirement, and minimal development effort. The design of PID controller is same for TCR and TCSC.

5. CONCLUSION

This paper presents the TCR and TCSC controller developed by using the MATLAB/Simulink. The developed software package consists of two main application menus which are the TCSC menu and the TCR-based SVC menu. These menus include seventeen simulation models about different applications of TCSC and TCR-based SVC. The effects of the TCSC and TCR-based SVC on load voltage have been studied in the single-phase and three-phase system with static load types. Besides, a single-machine infinite-bus system with static load type has been studied. The studied power systems are two and three bus with a long transmission line model. In this paper, we have demonstrated few applications of the FACTS controller such as a single-phase system with TCSC for the static load, and a SMIB system with TCR-based SVC for the static load. The simulation results show that significant improvement on voltage regulation and reactive power compensation is obtained by using the TSC and the TCR-based SVC. A survey which has six statements regarding facts controller was prepared. According to the survey, majority of the students thought that the MATLAB software package is user-friendly, easy to understand and several system parameters could be changed easily. This package is considerably effective for students and instructors to study theory of controlled compensators, the reactive power control and voltage regulation. Future work will concentrate on designing laboratory prototypes of the TCSC and the TCR-based SVC devices to provide the ability to experimentally verify the MATLAB software package.

REFERENCES

[1] NG Hingorani, L Gyugyi, Understanding FACTS, *IEEE Press*, New York, 1999.
[2] Aboytes F, Arroyo G, Villa G. Application of Static V AR Compensators in Longitudinal Power Systems, *IEEE Trans on PAS*, 1983; 102: 3460-3466.
[3] R Arulmozhiyal, K Baskaran, Implementation of a Fuzzy PI Controller for Speed Control of Induction Motors Using FPGA, *Journal of Power Electronics,* 2010; 10: 65-71.
[4] LAS Pilotto, AR Carvalho, A Bianco, WF Long, FL Alvarado, CL DeMarco, A Edris, *The Impact of Different TCSC Control Methodologies on the Subsynchronous Resonance Problem*, Proceedings of EPRI Conference on FACTS, Washington, DC, 1996.
[5] N Christl, P Luelzberger, M Pereira, K Sadek, PE Krause, AH Montoya, DR Torgerson, BA Vossler, *Advanced Series Compensation with Variable Impedance*, Proceedings of EPRI Conference on FACTS, 1990.
[6] RM Mathur, RK Varma, Thyristor-Based FACTS Controllers for Electrical Transmission Systems, *IEEE Press*, USA, 2002
[7] M Hedayati, *Technical specification and requirements of static VAr compensation (SVC) protection consist of TCR, TSC and combined TCR/TSC*, Proceedings of the 39th International Universities Power Engineering Conference, Bristol, UK, 2004; 1: 261–264.
[8] CIGRE Working Group 14.18, Thyristor-Controlled Series Compensation, *Technical Brochure*, Paris, 1996.
[9] A Ghosh, G Ledwich, *Modelling and control of thyristor controlled series compensators*, IEE Proc. Generation Transmission and Distribution, 1995; 142(3): 297-304.
[10] D Zhang, *et al.*, Common Mode Circulating Current Control of Interleaved Three-Phase Two-Level Voltage-Source Converters with Discontinuous Space-Vector Modulation, *2009 IEEE Energy Conversion Congress and Exposition,* 2009; 1(6): 3906-3912.
[11] A Gelen, T Yalcinoz, The behaviour of TSR-based SVC and TCR-based SVC installed in an infinite bus system, *Proceedings of the IEEE 25th Convention Electrical and Electronics Engineers in Israel*, Eilat, Israel, 2008; 120–124.
[12] Z Yinhai, *et al.*, A Novel SVPWM Modulation Scheme, *Applied Power Electronics Conference and Exposition, 2009. APEC 2009. Twenty-Fourth Annual IEEE*, 2009; 128-131.
[13] BK Johnson, *Simulation of TSC and TSSC*, Proceedings of the IEEE Power Engineering Society Winter Meeting, Columbus, USA, 2001; 637–63.

6

Selection of Power Semiconductor Switches in M.H.B.R.I. Fitted Induction Heater for Less Harmonic Injection in Power Line

Pradip Kumar Sadhu*, Palash Pal**, Nitai Pal*, Sourish Sanyal***

* Electrical Engineering Department, Indian School of Mines (under MHRD, Govt. of India), Dhanbad - 826004, India
** Department of Electrical Engineering, Saroj Mohan Institute of Technology (Degree Engineering Divison), a Unit of Techno India Group, Guptipara, Hooghly-712512, India
*** Department of Electronics and Communication Engineering, Academy of Technology, Hooghly, India

Keyword:

Eddy current
Induction heater
Modified half bridge resonant
Inverter
THD

ABSTRACT

This paper presents an approach to minimize the harmonics contained in input current of single phase Modified Half Bridge Resonant Inverter (M.H.B.R.I.) fitted induction heating equipment. A switch like IGBT, GTO and MOSFET are used for this purpose. It is analyzed the harmonics or noise content in the sinusoidal input current of this inverter. Fourier Transform has been used to distinguish between the fundamental and the harmonics, as it is a better investigative tool for an unknown signal in the frequency domain. An extensive method for the selection of different power semiconductor switches for Modified Half Bridge Resonant inverter fed induction heater is presented. Heating coil of the induction heater is made of litz wire which reduces the skin effect and proximity effect at high operating frequency. With the calculated optimum values of input current of the system at a particular operating frequency, the modified half bridge resonant inverter topology has been simulated using P-SIM software. From this proposed analysis the selection of suitable power semiconductor switches like IGBT, GTO and MOSFET are made. Waveforms have been shown to justify the feasibility for real implementation of single phase Modified Half Bridge Resonant inverter fed induction heater in domestic applications as well as industrial applications.

Corresponding Author:

Pradip Kumar Sadhu,
Department of Electrical Engineering,
Indian School of Mines (under MHRD, Govt. of India),
Dhanbad - 826004, India.
Email: pradip_sadhu@yahoo.co.in

1. INTRODUCTION

Induction heater for industrial applications [2] operates at a high frequency [4], [7] range from 1 kHz to 100 kHz. In the application of low frequency induction heating the temperature distribution can be controlled by slowly varying magnetic fields below a frequency [6] as low as 300 Hz. For medium frequency application, an auxiliary voltage-fed inverter is operated in parallel with the main current-fed inverter, since the current- fed parallel inverters [3] alone, when used for induction heating, fail to start. A high frequency modified half bridge inverters [1], [5] for induction heating and melting applications are self-started. For self-commutation, a resonant circuit is essential for SCR fitted inverter[8]-[9], [11]. It is assumed that the circuit is under damped; a mandatory condition for the circuit. The capacitor required for under damping can be connected in series or in parallel with the load [3] in the modern times, IGBTs [1], [8], GTOs and MOSFETs are preferred to SCRs mainly because they offer convenient turn OFF characteristics. Some auxiliary circuits and equipment are required to minimize switching losses occurring at high frequencies [2]. With the same designed parameters of the said inverter circuit, various switches such as IGBT, GTO and MOSFET have been used [10].

The requirements of induction heater are as follows:
a)   Switching in high frequency range
b)   High efficiency
c)   Power factor close to unity
d)   Wide power range and
e)   Reliability

Induction heaters are usually designed to operate with a vessel made from a specific material, mainly cast iron or ferro-magnetic stainless steel. The following is therefore desire characteristics for the inverter,

a)   No reactive components other than the heating coil and the non-smooth filter inductor,
b)   No input or matching transformer,
c)   50% duty ratio, simplifying the control and gate circuits,
d)   Clamped switch voltage and or current,
e)   The use of uncontrolled voltage source.

Here the complete inverter configuration has been simulated using P-SIM. In this present paper, response of harmonic injection in input power line of modified half bridge resonant inverter is tested & verified with different power switches and finally appropriateness of the switches is confirmed.

2.   ANALYSIS OF PROPOSED MODIFIED HALF BRIDGE RESONANT INVERTER

Proposed modified half bridge circuit is normally used for higher power output. Basic circuit is shown in the Figure 1. Four solid state switches are used and two switches are triggered simultaneously. Anti-parallel diodes are connected with the switch that allows the current to flow when the main switch is turned OFF. According to Figure 1, when there is no signal at Q_1 and Q_2, capacitors C_1 and C_2 are charged to a voltage of $V_i/2$ each. The Gate pulse appears at the gate of Q1 to turn IGBT ON. Capacitor C_1 discharges through the path NOPTN. At the same time capacitor C_2 charges through the path MNOPTSYM. The discharging current of C_1 and the charging current of C_2 simultaneously flow from P to T. In the next slit of the gate pulse, Q_1 and Q_2 remain OFF and the capacitors charge to a voltage $V_i/2$ each again. The Gate pulse appears at the gate of Q_2, so turning on Q_2. The capacitor C_2 discharges through the path TPQST and the charging path for capacitor C_1 is MNTPQSYM. The discharging current of C_2 and the charging current of C_1 simultaneously flow from T to P.

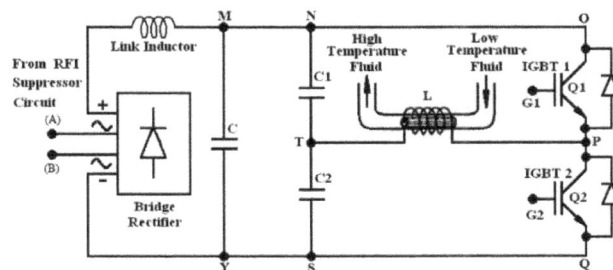

Figure 1. Proposed modified half bridge resonant inverter

Figure 2 indicates a specially designed eddy current heated metallic package which is tightly integrated into then on- metallic vessel or tank in the pipe line. The mechanically processed thin stainless-steel layer package with many spots and fluid channels for cylindrical induction-heated assembly is demonstrated in Figure 3.

When the fluid flows through the inherent package in the vessel or tank having a working coil connected to pipeline, the turbulent fluid is heated abruptly by eddy current losses generated inside the stainless-steel package. Internal structure of this metallic package to be heated by eddy current losses is indicated in Figure 3.

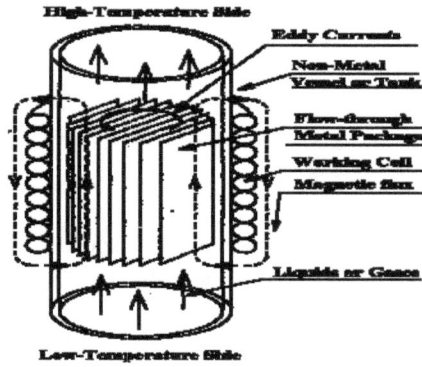

Figure 2. Heating package in the vessel and tank

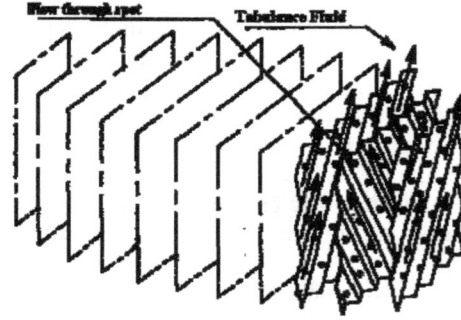

Figure 3. Internal structure of fluid through metal layer packing to generate turbulence flow

3.   CIRCUIT EQUATIONS

3.1.   Instantaneous Current i_0

With inductive load the equation of instantaneous current i_0 can be obtained as:

$$i_0(t) = \sum_{n=1,3,5,\ldots}^{\infty} \frac{2V_i}{n\pi\sqrt{R^2 + (n\omega L)^2}} \sin(n\omega t - \theta_n)$$

Here, $Z_n = \sqrt{R^2 + (n\omega L)^2}$ is the impedance offered by the load to the nth harmonic component, $\dfrac{2V_i}{n\pi}$ is the peak amplitude of nth harmonic voltage, and:

$$\theta_n = \tan^{-1}\left(\frac{n\omega L}{R}\right)$$

3.2.   Output Power

The output power at fundamental frequency (n=1) is given by:

$$P_{1_{rms}} = E_{1_{rms}}.I_{1_{rms}}.\cos\theta_1 = I_{1_{rms}}^{\ 2}.R$$

Where, $E_{1_{rms}}$ =RMS value of fundamental output voltage.

$I_{1_{rms}}$ =RMS value of fundamental output current.

$$\theta_1 = \tan^{-1}\left(\frac{\omega L}{R}\right)$$

But,

$$I_{1_{rms}} = \frac{2V_i}{\sqrt{2}.\pi.\sqrt{R^2 + (\omega L)^2}}$$

$$P_{1_{rms}} = I_{1_{rms}}^{\ 2}.R = \left[\frac{2V_i}{\pi.\sqrt{2}.\sqrt{R^2 + (\omega L)^2}}\right]^2.R$$

$$= \left[\frac{4V_i^2.R}{2\pi^2(R^2 + \omega^2 L^2)}\right] = \left[\frac{2V_i^2.R}{\pi^2(R^2 + \omega^2 L^2)}\right]$$

In high frequency heating application the fundamental power is more important, the output power due to fundamental current is generally the useful power and the power due to harmonic current is dissipated as heat.

4. THE HARMONIC CONTENT

The input current waveforms of an ideal inverter should be sinusoidal. But, in practice, the input current waveforms are non-sinusoidal. It contains harmonics. The existence of harmonics is visualized either in the time-domain or in the frequency domain easily. The availability of high speed power semiconductor devices has enabled us to reduce the harmonic contents in the input voltage significantly by switching techniques [6]. Total Harmonic Distortion (THD) is a measure of the closeness of a waveform with its fundamental component. The task of the design engineer is to reduce THD. It is accomplished by an LC Low Pass filter (LPF) as well as using most suitable high frequency semiconductor switch. LPF append at the input power supply terminal o f Modified Half Bridge Resonant Inverter for induction heating equipment. It provides low harmonic impedance to ground.

4.1. Analytical Tools

The quality of input current of a Modified Half Bridge Resonant Inverter i s obtained by Fast Fourier's analysis. It is a powerful mathematical tool which separates out the fundamental and the harmonics. Fourier's transforms allows us to peep into the frequency domain representation of the waveform.

4.1.1. Total Harmonic Distortion (THD)

It is a measure of distortion of a waveform. It is given by the following expression:

$$T.H.D = \frac{\sqrt{\sum_{n=2,3}^{\alpha} I_{nr.m.s}^{2}}}{I_{1r.m.s}} \tag{1}$$

It is the ratio of the RMS value of all non-fundamental frequency components to the RMS value of the fundamental. Our aim is to reduce to a minimum. For a rectangular wave: The value is very large. In quasi-rectangular form, the value is relatively less.

T.H.D Calculation from Software Simulation:

a) T.H.D Calculation of Modified Half Bridge Resonant Inverter using MOSFET. The R.M.S Value of Input Current, $I_4 = 2.21A$.

$$T.H.D = \frac{\sqrt{\sum_{n=2,3}^{\alpha} I_{nr.m.s}^{2}}}{I_{1r.m.s}}$$

$$= \frac{\sqrt{(1.16 \times 10^{-2})^2 + (1.07 \times 10^{-2})^2 + (4.63 \times 10^{-2})^2 + (1.06 \times 10^{-1})^2}}{3.13} A$$

$$= 3.73\%$$

T.H.D Calculation of Modified Half Bridge Resonant Inverter using GTO. The R.M.S Value of Input Current, $I_4 = 4.08A$.

$$T.H.D = \frac{\sqrt{\sum_{n=2,3}^{\alpha} I_{nr.m.s}^{2}}}{I_{1r.m.s}}$$

$$= \frac{\sqrt{(8.95 \times 10^{-2})^2 + (2.56 \times 10^{-2})^2 + (5.95 \times 10^{-2})^2}}{4.08} A$$

$$= 2.7\%$$

T.H.D Calculation of Modified Half Bridge Series Resonant Inverter using IGBT. The R.M.S Value of Input Current, $I_4 = 4.097A$.

$$T.H.D = \frac{\sqrt{\sum_{n=2,3}^{\alpha} I_{nr.m.s}^{2}}}{I_{1r.m.s}}$$

$$= \frac{\sqrt{(2.38\times10^{-2})^{2} + (2.81\times10^{-2})^{2} + (7.57\times10^{-3})^{2} + (7.76\times10^{-2})^{2}}}{4.097} A$$

$$= 2.1\%$$

4.1.2. Fast Fourier Transform (FFT) Analysis:

A Fast Fourier Transform (FFT) is an algorithm to compute the discrete Fourier transform (DFT) and it's inverse. It is a linear algorithm that can transform a time domain signal into its frequency domain equivalent and back. An FFT is a way to compute the same result more quickly. FFTs are of great importance to a wide variety of applications, from digital signal processing and solving partial differential equations to algorithms for quick multiplication of large integers. A better understanding of an unknown signal is obtained in the frequency domain. Peak noise in the input current of Modified Half Bridge Resonant Inverter using MOSFET, GTO and IGBT with LPF filter is determined by FFT analysis. The magnitudes of peak noises are given in the following Table 1.

Table 1. Noise response of different power semiconductor switches

Noise Signal	Magnitude of Peak Noise Current of Modified Half Bridge Resonant Inverter		
	MOSFET	GTO	IGBT
1st Noise	1.16×10^{-2}	8.95×10^{-2}	2.38×10^{-2}
2nd Noise	1.07×10^{-2}	2.56×10^{-2}	2.81×10^{-2}
3rd Noise	4.63×10^{-2}	5.95×10^{-2}	7.57×10^{-3}
4th Noise	1.06×10^{-1}	—	7.76×10^{-2}

4.1.3. LC-Low Pass Filter

An L-C low pass filter (LPF) allows waves of lower frequency to pass out more easily compared to the waves of higher frequency. While cascaded with an inverter, it is designed for such a cut-off frequency that the higher harmonics face more impedance and get reduced in magnitude.

5. SIMULATION AND RESULTS

In this paper, the proposed modified half bridge resonant inverter has been simulated using P-SIM with the help of equivalent parameters connected at the input of the induction heated system. Here from this topology the waveforms have been obtained using P-SIM software using different power semiconductor switches using IGBT, GTO and MOSFET from modified half bridge resonant inverter circuit and harmonics can be obtained. Figure 4 shows the simulation diagram of modified half bridge resonant inverter circuit using MOSFET with Low pass filter. Figure 5 shows the Power simulated wave-form of the input current of the modified half bridge resonant inverter using MOSFET Switch. Figure 6 shows the FFT waveform of input current using MOSFET switch. It may be noted that the harmonics are dominant using MOSFET switch. Figure 7 shows the circuit configuration for the modified half bridge resonant inverter using GTO. Figure 8 shows the wave-form of the input current for the modified half bridge resonant inverter with GTO Switch. Figure 9 shows the FFT for the same. It may be noted that here also the harmonics are dominant. Figure 10 shows the circuit configuration for the modified half bridge resonant inverter using IGBT. Figure 11 shows the wave-form of the input current for the proposed inverter using IGBT Switches. Figure 12 shows the FFT for the same. Here also the harmonics are dominant.

It is observed that from the power simulated wave-shapes and mathematical analysis the noise in the input current is much less after using the IGBT switches from the proposed topology. It is exposed from the FFT that the harmonic contents are almost absent in IGBT switch using P-SIM software apart from other power semiconductor switches.

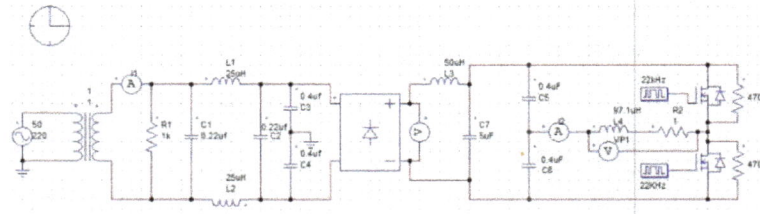

Figure 4. Modified Half Bridge Resonant Inverter circuit using MOSFET with LPF filter

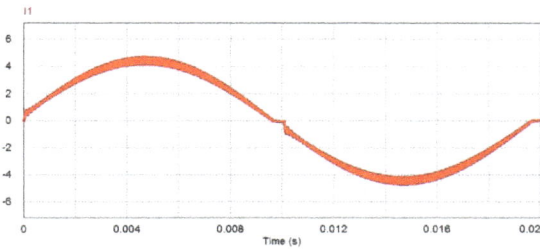

Figure 5. Input current waveform of Modified Half Bridge Resonant Inverter using MOSFET with LPF filter

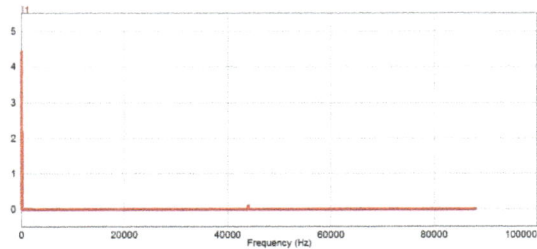

Figure 6. FFT of input current of the Modified Half Bridge Resonant Inverter using MOSFET with LPF filter

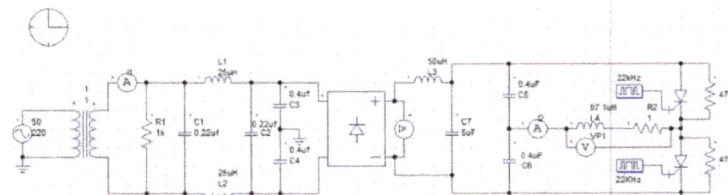

Figure 7. Modified Half Bridge Resonant Inverter circuit using GTO with LPF filter

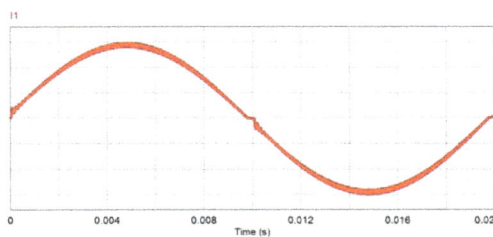

Figure 8. Input current waveform of Modified Half Bridge Resonant Inverter using GTO with LPF filter

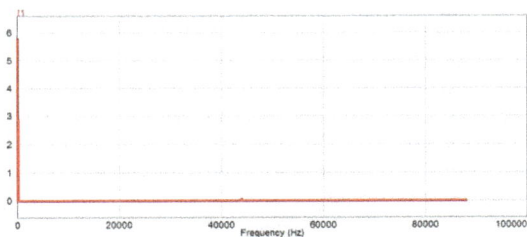

Figure 9. FFT of input current of the Modified Half Bridge Resonant Inverter using GTO with LPF filter

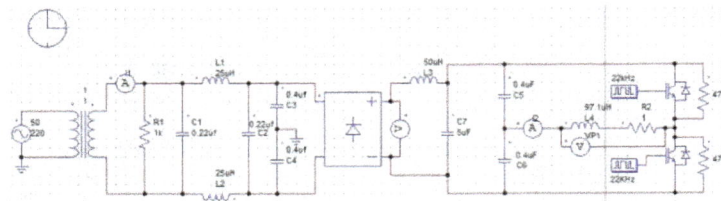

Figure 10. Modified Half Bridge Resonant Inverter circuit using IGBT with LPF filter

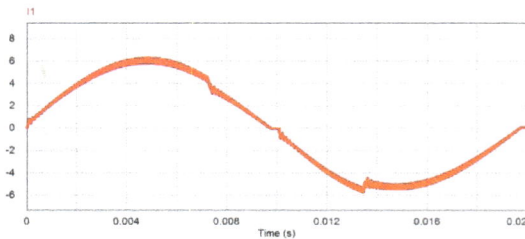

Figure 11. Input current waveform of Modified Half Bridge Resonant Inverter using IGBT with LPF filter

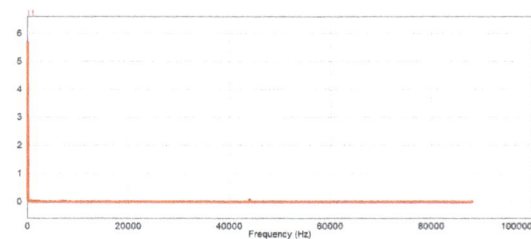

Figure 12. FFT of input current of the Modified Half Bridge Resonant Inverter using IGBT with LPF filter

6.   LABORATORY TEST BENCH

Figure 13. Photograph of Experimental Set-up

7.   CONCLUSION

Hence from the proposed topology it can be conclude that the different families of power semiconductor switches like GTO, IGBT and MOSFET are tested in modified half bridge resonant inverter fitted induction heater. To get minimum harmonics injection in the supply and to improve the efficiency of the inverter the proposed scheme can be employed in high frequency induction heating system. After comparing the wave-forms analysis of PSIM simulation, it is quite obvious that the selection of power semiconductor likely IGBT will be more suitable power semiconductor switch in high frequency modified half bridge resonant inverter. It has advantageous for reduced harmonic injection in power supply of induction heater. Again THD analysis is proven that selection of IGBT semiconductor switch is the best for induction heating applications.

ACKNOWLEDGEMENTS

Authors are thankful to the UNIVERSITY GRANTS COMMISSION, Bahadurshah Zafar Marg, New Delhi, India for granting financial support under Major Research Project entitled "Simulation of high-frequency mirror inverter for energy efficient induction heated cooking oven" and also grateful to the Under Secretary and Joint Secretary of UGC, India for their active co-operation.

REFERENCES

[1]   V Esteve, E Sanchis-Kilders, J Jordan, EJ Dede, C Cases, E Maset, JB Ejea, A Ferreres. Improving the efficiency of IGBT series-resonant inverters using pulse density modulation. *IEEE Trans. Ind.Electron.* 2011; 58(3): 979–987.

[2]   J Acero, JM Burdio, LA Barragan, D Navarro, R Alonso, J Ramon, F Monterde, P Hernandez, S Llorente, I Garde. Domestic induction appliances: An overview of recent research. *IEEE Ind. Appl. Mag.* 2010; 16(2): 39–47.

[3]   S Okudaira, K Matsuse. Adjustable frequency quasi-resonant inverter circuits having short-circuitswitch across resonant capacitor. *IEEE Transactions on Power Electronics.* 2008; 23(4): 1830–1838.

[4]   S Johnson, R Zane. Custom spectral shaping for EMI reduction in high frequency inverters and ballasts. *IEEE Trans. Power Electron.* 2005; 20: 1499–1505.

[5]   JM Burdio, F Monterde, JR Garcia, LA Barragan, A Martinez. A two-output series resonant inverter for induction-heating cooking appliances. *IEEE Transactions on Power Electronics.* 2005; 20: 815-822.

[6]   S Llorente, F Monterde, JM Burdio, J Acero. *A comparative study of resonant inverter topologies used in induction cooker.* IEEE Applied Power Electronics Conf. (APEC). 2002; 1168-1174.

[7] K Itoh, Y Moriura, T Satoh, K Arimatsu, N Nakayama, K Kimoto, T Doizaki, K Dojoh. *9000kW-1500 Hz frequency converter for hot bar heater*. Fourth Power Conversion Conference-NAGOYA, PCC-NAGOYA 2007 - Conference Proceedings. 2007; 904–910.

[8] Shimada, JA Wiik, T Isobe, T Takaku, N Iwamuro, Y Uchida, M Molinas, TM Undeland. *A new AC current switch called MERS with low on-state voltage IGBTs (1.54 V) for renewable energyand power saving applications*. Proceedings of the 20th International Symposium on Power Semiconductor Devices & IC's. 2008; 4–11.

[9] PK Sadhu, RN Chakrabarti, SP Chowdhury. An improved inverter circuit arrangement. Patent No. 69/Cal/2001. Patent Office – Government of India.

[10] Pradip Kumar Sadhu, Debabrata Roy, Nitai Pal, Sourish Sanyal. Selection of Appropriate Semiconductor Switches for Induction Heated Pipe-Line using High Frequency Full Bridge Inverter. *International Journal of Power Electronics and Drive System (IJPEDS, SCIMago Journal)*. 2014; 5(1): 112–118.

[11] Mochammad Facta, Tole Sutikno, Zainal Salam. The Application of FPGA in PWM Controlled Resonant Converter for an Ozone Generator. *International Journal of Power Electronics and Drive System (IJPEDS, SCIMago Journal)*. 2013; 3(3): 336~343.

7

Power Quality Improvement Using Custom Power Devices in Squirrel Cage Induction Generator Wind Farm to Weak-Grid Connection by using Neuro-fuzzy Control

Kopella Sai Teja, R.B. R.prakash
Departement of Electrical and Electronics Engineering, K L University

ABSTRACT

Keyword:

DClink
Neuro-fuzzy logic control
Simulation
SCIG
UPQC

Wind farm is connected to the grid directly.The wind is not constant voltage fluctuations occur at point of common coupling (PCC) and WF terminal. To overcome this problem a new compensation strategy is used. By using Custom power devices (UPQC).It injects reactive power at PCC. The advantages of UPQC are it consists of both DVR and D-STATCOM. DVR is connected in series to the line and it injects in phase voltage into the line .D-STATCOM is connected shunt to the line .The internal control strategy is based on management of active and reactive power in series and shunt converters of UPQC. The power exchainge is done by using DC-link.

Corresponding Author:

Kopella sai teja,
Departement of Electrical and Elctronics Engineering,
K L University.
Email: kopellasaiteja@live.com

1. INTRODUCTION

The location of generation facilities for wind energy is determined by wind energy resource availability, often far from high voltage (HV) power transmission grids and major consumption centres. In case of facilities with medium power ratings, the Wind Farm is connected through medium voltage (MV) distribution headlines.

Also, is well known that given the random nature of wind resources, the wind farm generates fluctuating electric power. These oscillations have a negative impact on stability and power quality in electric power systems. furthermore, in development of wind resources, turbines utilizing squirrel cage induction generators (SCIG) have been used since the beginnings. The operation of squirrel cage induction generator demands reactive power, generally provided from the mains and/or by local generation in capacitor banks [1].

In the event that changes occur in its mechanical speed, i.e. due to wind disturbances will the WIND FARM active (reactive) power injected (demanded) into the power grid, leading to variations of wind farm Terminal voltage because of system impedance. This power disturbances transmit into the power system, and can produce a phenomenon known as "flicker", which consists of fluctuations in the illumination level caused by voltage variations. Also, the normal operation of Wind Farm is impaired due to such disturbances. In particular for the case of "weak grids", the impact is even better.

In order to reduce the voltage fluctuations that may cause "flicker", and improve Wind Farm terminal voltage regulation, several results have been posed. The most common one is to raise the power grid, enhancing the short circuit power level at the point of common coupling point of common coupling, thus reducing the impact of power fluctuations and voltage regulation problems.

In recent years, the technological development of high power electronics devices has led to implementation of electronic equipment suited for electric power systems, with very fast response compared to the line frequency. These active compensators allow groovy flexibility in: I) controlling the power flow in transmission systems using Flexible AC Transmission System (FACTS) devices, and II) enhancing the power quality in distribution systems employing Custom Power System (CUPS) devices [2]. The use of these active compensators to improve integration of wind energy in weak grids is the approach adopted in this work. In this project we analyse a compensation strategy using an UPQC, for the SCIG–based Wind Farm connected to a weak distribution power grid. This system is taken from a real case [3].

The UPQC is controlled to regulate the Wind Farm terminal voltage, and to mitigate harmonics at the point of common coupling (PCC), caused by system load changes in generated power of Wind Farm, respectively. By using UPQC series converter in wind farm voltage regulation process was done, by voltage injection "in phase" with PCC voltage.

The shunt converter is used to filter the Wind Farm generated power to forbid voltage flickers in active and reactive power capability. The sharing of active power between converters, is supervised through the common DC link.

2.   SYSTEM DESCRIPTION AND MODELLING
2.1. System Description

Figure 1 depicts the power system under consideration in this study.

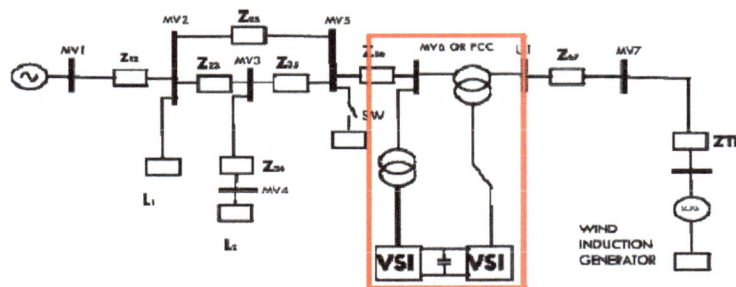

Figure 1. Single line diagram of wind farm connected to week grid system

The Wind Farm is composed by 36 wind turbines using SCIG, adding up to 21.6MW electric power. Each turbine has given fixed reactive compensation capacitor banks (175kVAr), and is connected to the power grid via 630KVA 0.69/33kV transformer. This system is Carry out from, and represents a real case.

The ratio between short circuit power and rated WIND FARM power, give us an idea of the connection weakness. Thus considering that the value of short circuit power in MV6 is SSC ≃ 120MV A this ratio can be calculated:

$$r = \frac{S_{SC}}{P_{WF}} \approx 5.5$$

Values of r < 19 are considered as a "weak grid" connection [2].

2.2. Turbine Rotor and Associated Disturbances Model

The power that can be obtained from a wind turbine, is expressed by:

$$P = \frac{1}{2}\rho\pi R^2 V^3 C_p$$

Where: ρ is air density

R the radius of the swept area
v the wind speed
CP the power coefficient

For the assumed turbines (600kW) the values are R = 31.2 m , ρ = 1.225 kg/m3 and CP calculation is taken from [4].

A complete model of the wind farm is obtained by turbine aggregation; this means that the whole wind farm can be modelled by only a equivalent wind turbine, whose power generated by the arithmetic sum of each turbine according to the equation given below:

$$P_T = \sum_{i=1 \ldots 36} P_i$$

Wind speed v in eqn (1) can differ around its average value due to variation in the wind flow. Such variations can be classified as random and deterministic. The first are caused by the symmetry in the wind flow observed by the turbine blades due to tower shadow and due to the atmospheric boundary layer, while the latter are random changes known as turbulence. For our analysis, wind flow variation due to support structure is considered, and modeled by a sinusoidal modulation superimposed to the mean value of v. The frequency for this modulation is $3.N_{rotor}$ for the three bladed wind turbine, while its distance depends on the geometry of the tower. In our case we have considered a mean wind speed of 12m/s and the amplitude modulation of 15%.

The effect of the boundary layer can be ignored compared to those produced by the shadow effect of the tower in almost cases [3]. It should be noted that while the arithmetic sum of perturbations occurs when all turbines function synchronously and in phase, this is the case that has the great impact on the power grid, since the power pulsation has high amplitude. So, turbine aggregation method is valid.

The effect of the boundary layer can be ignored compared to those produced by the shadow effect of the tower in almost cases [3]. It should be noted that while the arithmetic sum of perturbations occurs when all turbines function synchonously and in phase, this is the case that has the great impact on the power grid, since the power pulsation has high amplitude. So, turbine aggregation method is valid.

3.   MODEL OF INDUCTION GENERATOR

The model available in Matlab/Simulink Sim Power Systems library the squirrel cage induction generator is used. It consists of a second–order mechanical model and a fourth–order state–space electrical model [5].

4.   DYNAMIC COMPENSATOR MODEL

The dynamic compensation of voltage changes is performed by injecting voltage in series and active & reactive power into the MV6 (PCC) busbar; this is accomplished by using an UPQC [1]. In Figure 2 we can see the basic single line diagram of this compensator; the impedances and busbars numbering is referred to Figure 1.

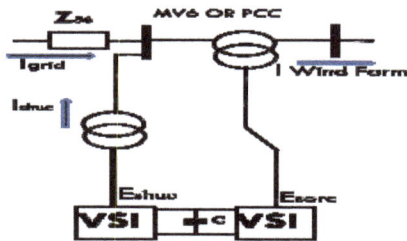

Figure 2(a). Block diagram of UPQC   Figure 2(b). Phasor diagram of UPQC

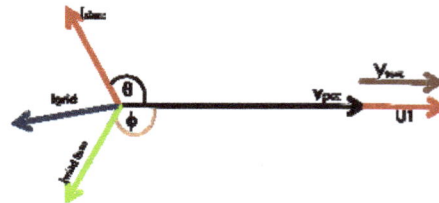

The operation is based on the generation of three phase voltages, using power electronic converters either current source type Current Source Inverter or voltage source type Voltage Source Inverter. Voltage Source converters are preferred. Faster response in the system than CSI [1] and It has lower DC link losses. The shunt converter of UPQC injecting current at PCC, hear the series converter generates voltages between U1 and PCC, illustrated in the phasor diagram of Figure 3. An important feature of this compensator is the operation of both VSI converters sharing the same DC–bus, it enables the active power exchange between them.

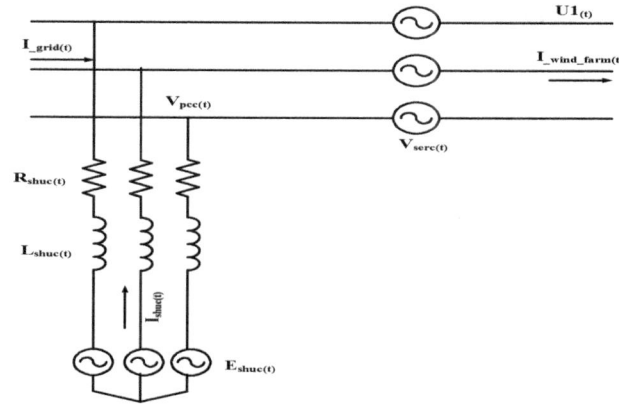

Figure 3. Power stage compensation model AC side

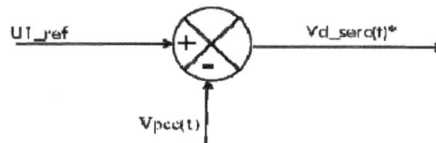

Figure 4. Series compensator controller

We have build up a simulation model for the UPQC based on the ideas chosen from [6]. Since switching control of converters is complete different of this work, and considering that higher order harmonics generated by VSI converters are outside the bandwidth of significance in the simulation study, the converters are modelled using ideal controlled voltage sources. Figure 4 shows the adopted model of power side of UPQC.

The control of the UPQC, will be enforced in a rotating frame dq0 using Park's transformation as given in Equation (3&4).

$$T = \frac{2}{3} \begin{bmatrix} \sin(\theta) & \sin(\theta - \frac{2\pi}{3}) & \sin(\theta + \frac{2\pi}{3}) \\ \cos(\theta) & \cos(\theta - \frac{2\pi}{3}) & \cos(\theta + \frac{2\pi}{3}) \\ \frac{1}{2} & \frac{1}{2} & \frac{1}{2} \end{bmatrix}$$

$$\begin{bmatrix} f_d \\ f_q \\ f_0 \end{bmatrix} = T . \begin{bmatrix} f_a \\ f_b \\ f_c \end{bmatrix}$$

Where: $f_i = a,b,c$ represents phase voltage or currents

   $f_i = d,q,0$ represents magnitudes transformed to the dqo space.

This transformation admits the alignment of a rotating reference frame with the positive sequence of the PCC voltages space vector. To attain this, a reference angle_synchronized with the PCC positive sequence fundamental voltage space vector is calculated using a Phase Locked Loop (PLL) system. In this work, an "instantaneous power theory" based PLL has been enforced [7].

Under balance steady-state conditions, voltage and currents vectors in this synchronous reference frame are constant quantities. This strategy is useful for analysis and decoupled control.

5.   UPQC CONTROL STRATEGY

In this paper we have used the neuro – fuzzy logic controlling strategy which is very advanced strategy now a days.

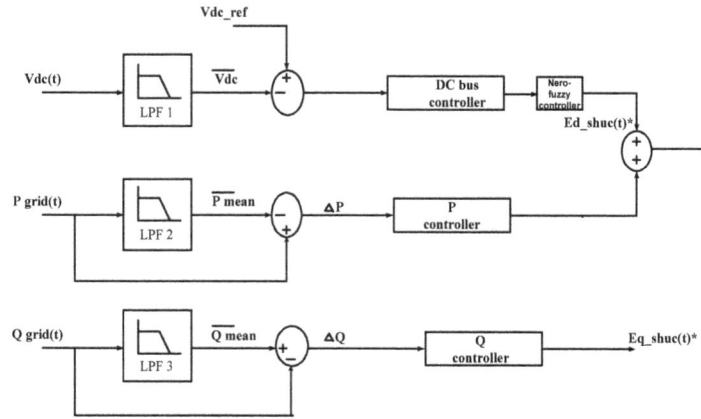

Figure 5. Shunt compensator controller using neuro – fuzzy

The powers P_{shuc} and Q_{shuc} are calculated in the rotating reference frame, as follow:

$$P_{shuc}(t) = \frac{3}{2}.V_d^{pcc}(t).I_d^{shuc}(t)$$

$$Q_{shuc}(t) = -\frac{3}{2}.V_d^{pcc}(t).I_q^{shuc}(t)$$

We Ignore PCC voltage variation, the above equations can be written as follows:

$$P_{shuc}(t) = k_p'.I_{d\_shuc}(t)$$

$$Q_{shuc}(t) = k_q'.I_{q\_shuc}(t)$$

Taking in consideration that the shunt converter is based on a VSI, we need to generate adecuate voltages to obtain the currents in equation. This is attained using the VSI model proposed in [6], leading to a linear relationship between the controller voltages and generated power. The resultant equations are:

$$P_{shuc}(t) = k_p''.E_{d\_shuc}^*(t)$$

$$Q_{shuc}(t) = k_q''.E_{q\_shuc}^*(t)$$

P and Q control loops comprise a PI controller, while DC–bus loop hear we use a neuro – fuzzy controller, In generally, in the proposed scheme the UPQC can be seen as a *power buffer*, leveling the power injected into the power system grid. The Figure 7 illustrates a conceptual diagram of this mode of operation.

It must be observed that the absence of an external DC source in the UPQC bus, forces to maintain zero–average power in the storage element installed in that bus. This is accomplished by a proper design of DC voltage controller.

Also, it is necessary to note that the proposed scheme cannot be implemented using other CUPS devices like DVR or D–Statcom. The power buffer concept may be implemented using a DStatcom, but not using a DVR. On the other side, voltage regulation during relatively large disturbances, cannot be easily using reactive power only from DStatcom; in this work, a DVR device is more suitable.

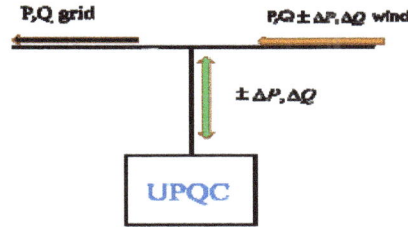

Figure 6. Power buffer concept

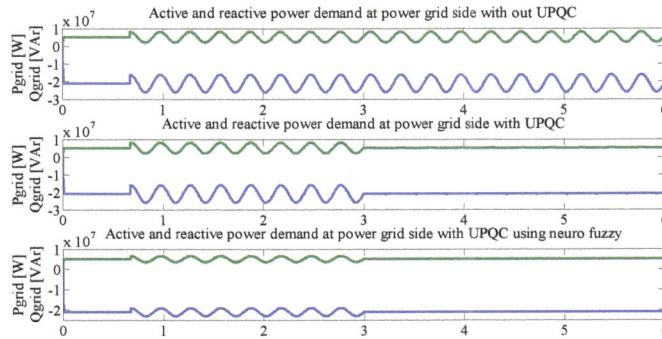

Figure 7. Active and reactive power demand at power grid

The model of the power system strategy illustrated in Figure 1, including the controllers with the control scheme detailed in section III, was implemented using Matlab/Simulink ® software. Numerical simulations were performed to determine and then compensate voltage fluctuation due to wind power variation, and voltage regulation problems due to a sudden load connection. The simulation was conducted with the following chronology.

   a) At t = 0.0'' the simulation starts with the series converter and the DC-bus voltage controllers in operation.
   b) At t = 0.5'' the tower shadow effect starts
   c) At t = 3.0" Q and P control loops
   d) At t = 6.0" L3 load is connected.
   e) At t = 6.0 " L3 load is disconnected[8].

6.   COMPENSATION OF VOLTAGE FLUCTUATION

Simulation results for 0 < t < 6 are shown in Fig.8. At t = 0.5″ begins the cyclical power pulsation produced by the tower shadow effect. As was mentioned, the tower shadow produces variation in torque, and hence in the active and reactive wind farm generated power. For nominal wind speed condition, the power fluctuation frequency is f = 3.4Hz, and the amplitude of the resulting voltage variation at PCC, expressed as a percentage is:

$$\frac{\Delta U}{U_{rated}} = 1.50\%$$

7.   RESULTS AND ANALYSIS

Figure 8 is the pcc voltage is behaviour is shown the upper curve shows the voltage at pcc when UPQC is not existing. The middle curve shows the when the UPQC is connected to the grid by using PI controller. The last curve in Figure 8 shows when UPQC is connected to the grid and hear we are using neuro –fuzzy logic controller. There is a variation in the wave forms.

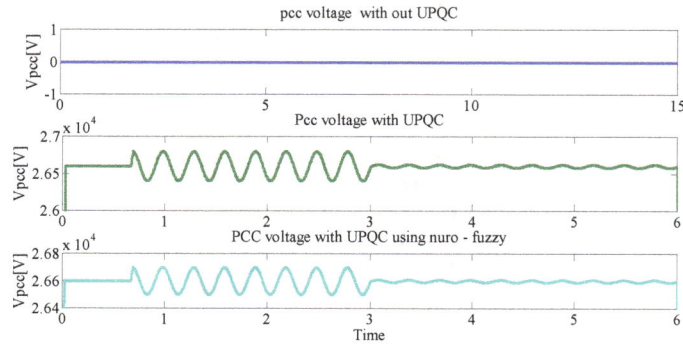

Figure 8. Pcc voltage

In the Figure 9 the shows the behaviour of wind farm terminal voltage. The upper wave form shows the behaviour when wind farm is connected to the grid when UPQC is not connected. Middle wave form is when UPQC is connected to the grid and the control strategy used is PI controller. Final wave form of Figure 9 shows the behaviour of wind farm voltage connected to the grid and the control strategy used is neuro – fuzzy logic controller.

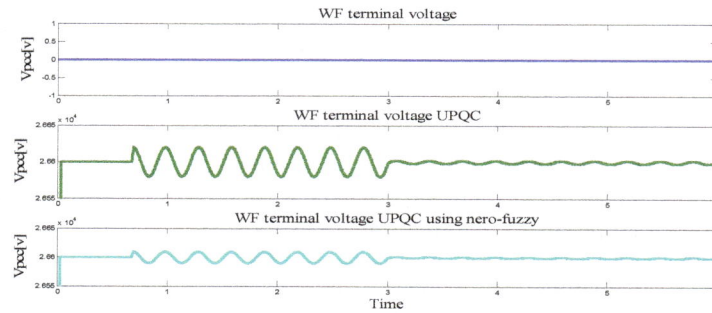

Figure 9. WF terminal voltage

This voltage fluctuation is seen in Figure 9 for 0.5 < t < 3.

Table 1. Neuro Fuzzy Rule Base

E(K) ΔE	NB	NM	NS	ZE	PS	PM	PB
NB	NB	NB	NB	NB	NM	NS	ZE
NM	NB	NB	NB	NM	NS	ZE	PS
NS	NB	NB	NM	NS	ZE	PS	PM
ZE	NB	NM	NS	ZE	PS	PM	PB
PS	NM	NS	ZE	PS	PM	PB	PB
PM	NS	ZE	PS	PM	PB	PB	PB
PB	ZE	PS	PM	PB	PB	PB	PB

In this paper the above rules are taken for [9]. In the above table PB=positive big, PM = positive medium, PS = positive small ZE = zero, NS = negative small, NM = negative medium, NB = negative big

In the Figure 10 the shows the behaviour of wind farm terminal voltage and Pcc voltage. The upper wave form shows the behaviour when wind farm is connected to the grid when UPQC is not connected.

middle wave form is when UPQC is connected to the grid and the control strategy used is PI controller. Final wave form of Figure 10 shows the behaviour of wind farm voltage connected to the grid and the control strategy used is neuro – fuzzy logic controller.

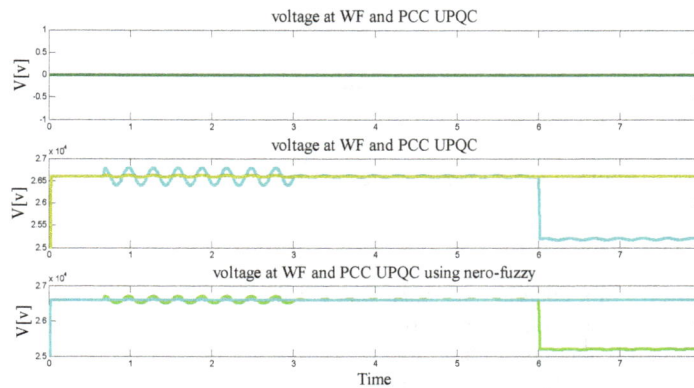

Figure 10. Voltage at Pcc and WF

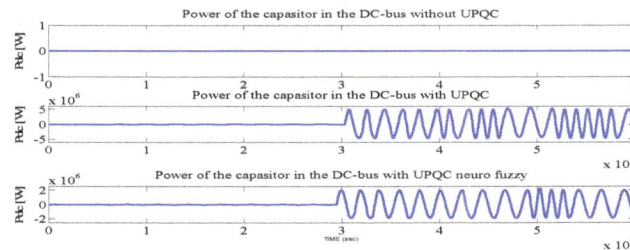

Figure 11. Power of capacitor in DC bus

The above Figure 11 is DC bus voltage at UPQC, in upper wave form there is no UPQC is connected to the grid. There is no UPQC no DC bus voltage so it is 0. The middle wave form is UPQC is connected to the grid there is variation. We have taken variation from time interval 3, so the variation starts from 3. The final wave form is also same hear we use neuro fuzzy.

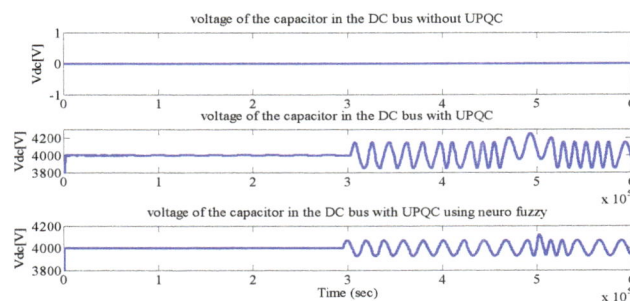

Figure 12. Voltage of the capacitor in the dc bus

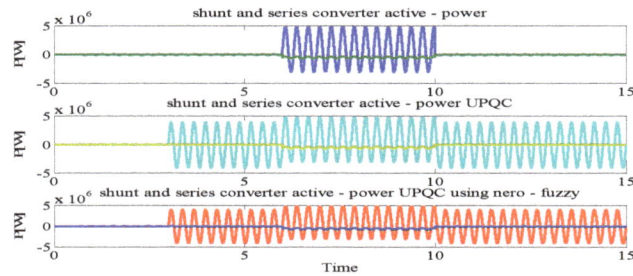

Figure 13. Shunt and series converter active power

8. CONCLUSION

In this paper, a new compensation strategy was employed using UPQC compensator. When SCIG based wind farms connected to week grid this compensation strategy is used. This compensation strategy enhances system power quality. The simulation results show the good performance in mitigating the power fluctuations due to tower shadow effect and voltage regulation in sudden load conditions.

REFERENCES

[1] A Ghosh, G Ledwich. Power Quality Enhancement Using Custom Power Devices. Kluwer Academic Publisher.

[2] Z Saad-Saoud, ML Lisboa, JB Ekanayake, N Jenkins and G Strbac. *Application of STATCOM's to wind farms.* IEE Proc. Gen. Trans. Distrib. 1998; 145(5).

[3] P Rosas. Dynamic influences of wind power on the power system. *Technical report RISØR-1408.* Ørsted Institute. March 2003.

[4] T Burton, D Sharpe, N Jenkins, E Bossanyi. Wind Energy Handbook. John Wiley & Sons, 2001. ISBN 0-471-48997-2.

[5] P Kundur. Power System Stability and Control. McGraw-Hill, 1994. ISBN 0-07-035958-X

[6] C Schauder, H Mehta. *Vector analysis and control of advanced static VAR compensators.* IEE PROCEEDINGS-C. 1993; 140(4).

[7] MF Farias, PE Battaiotto, MG Cendoya. Wind Farm to Weak-Grid Connection using UPQC Custom Power Device, *IEEE.* 2010

[8] M Vishnu vardhan, Dr P Sangameswararaju.Using NFC and modified CS Algorithm based unified power flow conditioner for compensating power quality problem. *International Journal of Scientific & Engineering Research.* 2013; 4(9).

8

Power Quality Improvement in Distribution System using ANN Based Shunt Active Power Filter

Jarupula Somlal, Venu Gopala Rao.Mannam, Narsimha Rao.Vutlapalli
Departement of Electrical and Electronics Engineering, K L University, Guntur, INDIA

Keyword:

Distribution system
Nueral network controller
Shunt active power filter
Total harmonic distortion

ABSTRACT

This paper focuses on an Artificial Neural Network (ANN) controller based Shunt Active Power Filter (SAPF) for mitigating the harmonics of the distribution system. To increase the performance of the conventional controller and take advantage of smart controllers, a feed forward-type (trained by a back propagation algorithm) ANN-based technique is implemented in shunt active power filters for producing the controlled pulses required for IGBT inverter. The proposed approach mainly work on the principle of capacitor energy to maintain the DC link voltage of a shunt connected filter and thus reduces the transient response time when there is abrupt variation in the load. The entire power system block set model of the proposed scheme has been developed in MATLAB environment. Simulations are carried out by using MATLAB, it is noticed that the %THD is reduced to 2.27% from 29.71% by ANN controlled filter. The simulated experimental results also show that the novel control method is not only easy to be computed and implemented, but also very successful in reducing harmonics.

Corresponding Author:

Jarupula Somlal,
Departement of Electrical and Electronics Engineering,
K L University,
Green Fields, Vaddeswaram, Guntur District, Andhra Pradesh, INDIA.
Email: jarupulasomu@kluniversity.in

1. INTRODUCTION

In recent years with the expansion of power semiconductor technology, power electronics based devices such as adjustable-speed drives, arc furnace, switched-mode power supply, uninterruptible power supply etc are employed in various fields [1], [5]-[7]. Some of these converters not only increase reactive currents, but also produce harmonics in the source current. Due to the harmonics, there many losses in the power system. To mitigate the harmonics, there are different solutions are proposed and used by researchers in literature such as line conditioners, passive filters, active filter, etc., Firstly, conventional passive filter are used for elimination of the harmonics; but these passive filter having some disadvantages; such as large in size ,fixed harmonic compensation, weight and resonance occurrence. The above drawbacks of passive filter can be overcome by the concept of active power filter approach. Shunt-type active power filter (SAPF) is used to eliminate the current harmonics. The SAPF topology is connected in parallel for current harmonic compensation. The shunt active power filter has the capability to maintain the mains current balanced and sinusoidal after compensation regardless of whether the load is non-linear and unbalanced or balanced. Recent technological developments of switching devices and availability of inexpensive controlling devices, e.g., DSP-field-programmable-gate-array-centered system, accomplish an active power line conditioner, a natural option to compensate for harmonics. The controller is the heart or primary component of the SAPF system. Conventional PI and PID controllers are used to extract the fundamental component of the load current thus facilitating reduction of harmonics and simultaneously controlling dc-side capacitor voltage of the voltage source inverter. Recently, different AI techniques controllers are used for shunt active power filters.

The major research works are related with control circuit design .The target is to obtain reliability control algorithms of the reference current and a quick response procedure to get the control signal and simultaneously quick controlling dc-side capacitor voltage of the voltage source inverter. The Artificial Neural Networks (ANNs) have been systematically applied to electrical engineering [2-3]. This method is considered as a new tool to design SAPF control circuits. The ANN presents two principal characteristics .It's not necessary to establish specific input-output relationships but they are formulated through a learning process. Moreover, the parallel computing architecture increases the system speed and reliability [4].

In this paper, a new SAPF control method based on ANNs will be presented. Load voltages and currents are sensed, the control blocks calculates the power circuit control signals from the reference compensation currents , and the power circuit injects the compensation current to power system. The article is primary focused on a system which uses the ANN system and the results for the same are discussed. In this paper, a shunt APF with a hysteresis band control is utilized to compensate the non-linear loads.

2. CONFIGURATION OF SHUNT ACTIVE POWER FILTER(SAPF) AND ESTIMATION OF COMPENSATING CURRENT

Figure 1 shows a shunt active power filter, it consists of the 3-phase source, universal bridge, load along with active filters. A SAPF is to produce the compensation current. The non-linear load is the sum of source current and the harmonic current.The objective is to get the balanced supply current with out harmonic and reactive components.The suitable current is injected by the SAPF corresponding to the load current. The SAPF is designed with ANN controller. The proposed controller, accounts for THD and DC voltage control, the controller have rapid dynamic response in case of load current deviation. The proper operation of the controller results in the generation of gate signals for 3-phase inverter which in turn is responsible for generating compensating currents. These compensating currents on injection through the 3-phase inverter results in harmonic compensation of source currents and improvement of power quality on the connected power system [9].

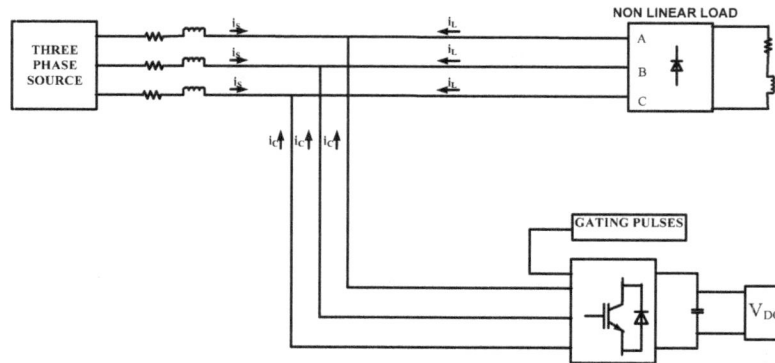

Figure 1. Configuration of Shunt Active Power Filter

A general formulation for the load current corresponding to Figure 1 is:

$$i_L(t) = i_{\alpha 1}(t) + i_{\beta 1}(t) + i_h(t) \tag{1}$$

$i_{\alpha 1}$ and $i_{\beta 1}$ are the in-phase and quadrature components of the phase current at the fundamental frequency respectively. All other harmonics are included in i_h. The per-phase source voltage and the corresponding in-phase component of the load current may be conveyed as:

$$v_s(t) = V_m \cos \omega t \tag{2}$$

$$i_{\alpha 1}(t) = I_{\alpha 1} \cos \omega t \tag{3}$$

Assuming that harmonics can be eliminated by the APF, the compensating current becomes:

$$i_L(t) = i_L(t) - i_{\alpha 1}(t) = i_L(t) - I_{\alpha 1} \cos \omega t \qquad (4)$$

Where $i_{\alpha 1}$ is the peak magnitude of the in-phase current that the mains should supply and therefore needs to be assessed Once $i_{\alpha 1}$ assessment is over, the reference current for the active power filter may easily be fixed as per (4). i_L may be measured using current sensors.

3. PROPOSED CONTROL STRATEGIES
3.1. Reference Current Calculation:

For reference current calculation, instantaneous abc_to_dq0 transformation has been applied. The abc_to_dq0 transformation block calculates the d-axis, q-axis, and zero sequence quantities in a two-axis rotating reference frame for a 3-Φ sinusoidal signal. Equation (5), (6) and (7) are used for reference current calculation,

$$I_d = \frac{2}{3}(I_a \sin(\omega t) + I_b \sin(\omega t - \frac{2\Pi}{3}) + I_C \sin(\omega t + \frac{2\Pi}{3})) \qquad (5)$$

$$I_q = \frac{2}{3}(I_a \cos(\omega t) + I_b \cos(\omega t - \frac{2\Pi}{3}) + I_C \cos(\omega t + \frac{2\Pi}{3})) \qquad (6)$$

$$I_o = \frac{1}{3}(I_a + I_b + I_c) \qquad (7)$$

Where ω = rotation speed (rad/s) of the rotating frame

3.2. Design of ANN Controller

An Artificial neural network (ANN), is a model (mathematical) inspired by biological neural networks. An ANN consists of an interlinked collection of artificial neurons, and it develops information using a connectionist method to calculation. It resembles the brain in two facets: 1) The data is accumulated by the network through the learning process and, 2) Interneuron connection strengths are employed to store the data. These networks are categorized by their topology, the manner in which they communicate with their surroundings, the manner in which they are guided, and their capability to process information. ANNs are applied to solve artificial intelligence problems without necessarily creating a model of a real dynamic system.

The rapid spotting of the disturbance signal with high accuracy, fast processing of the reference signal, and high dynamic response of the controller are the prime prerequisites for desired compensation in case of APF. The conventional controller fails to achieve satisfactorily under parameter variations nonlinearity load disturbance, and so forth

For improving the performance of the suggested Shunt Active Power filter, single layer feed forward network (trained by the back propagation algorithm) is seen. This network consists of two layers and their corresponding neuron interconnections. '2' neurons in input layer to receive the inputs. Hidden layer comprises of 21 neurons to which each of the processed input is fed. The output layer comprises of '1' neuron whose output is to be calculated as P_{loss}. Activation functions are assigned for each of the layers in order to train them. Input layer is given the Tan-Sigmoidal function as activation function and the output layer is being given the Pos-Linear activation function as activation function.

Figure 2 shows the internal blocks of proposed neural network [10]. The large data of the DC-link voltage for 'n' and 'n-1' intervals from the conventional method are gathered and are stored in the MATLAB workspace. This data is used for training the ANN. The data stored in workspace is being retrieved using the training algorithm used. The neurons in the input and output layers is almost a fixed quantity to obtain the provided input. The accuracy of the ANN operation is mostly depends on the number of hidden neurons.

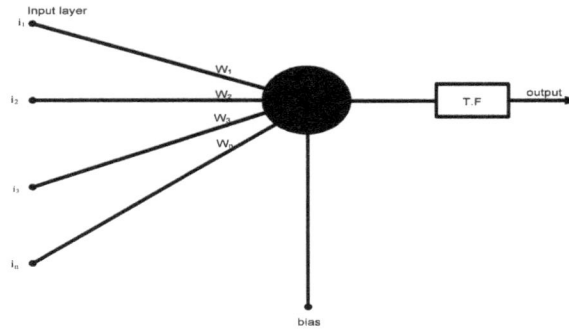

Figure 2. Internal blocks of proposed neural network

$$y = \sum_{n=1}^{21} w_n \cdot i_n + b \qquad\qquad (8)$$

3.2.1. Algorithm for ANN

Step 1: Normalize the inputs and outputs with respect to their maximum values. It is shown that the neural networks work better if the inputs and outputs lie between 0-1. There are two inputs given by $\{P\}_{2X20}$ and one output $\{O\}_{1X20}$ in a normalized form.

Step 2: Enter the number of inputs for a fed network.

Step 3: Enter the number of layers.

Step 4: Create a new feed forward network with 'tansig and poslin' transfer functions.

Step 5: Train the network with a learning rate 0.02.

Step 6: Enter the number of epochs.

Step 7: Enter the goal.

Step 8: Train the network for given input and targeted output.

Step 9: Generate simulation of the given network with a command 'gensim'

The Neural Network is created with the set number of neurons in the each layer using the above algorithm. At each training session, 500 iterations are done and 6 such a validation checks are taken out in order to minimize the scope of error occurrence. The main aim of this is to bring the performance to zero. The Learning rate is the major consideration in the training of the Artificial Neural Network (change of interconnection weights). It should not be too low that the training gets too delayed. It should not be excessively because the oscillations occur about the target values and the time needed to converge is too high and the training gets delayed. For the considered controller, Neural Network is trained at a learning rate of 0.02. The compensator output depends on the input and its evolution.

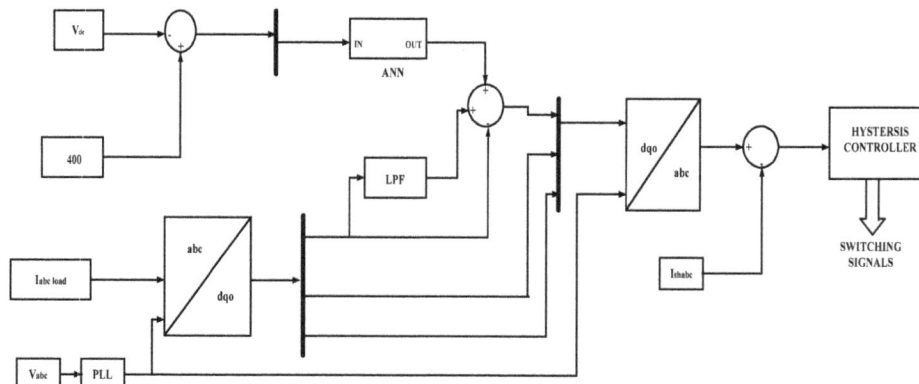

Figure 3. Control scheme for ANN controller

Figure 3 shows proposed control scheme for ANN, in which the load currents, PCC voltages and DC bus voltage of shunt active filter are sensed. The constant DC voltage is maintained by the DC voltage loop. The input of ANN controller is the difference between V_{DC} and a reference value. The output of ANN is responsible for harmonic mitigation. A phase-locked loop (PLL) synchronizes on the positive-sequence component of the current I. The output of the PLL (angle $\theta = \omega t$) is used to compute the direct-axis and quadrature-axis components of the three-phase currents [11]. The output signals of ANN controller and direct axis component of current from d-q-o transformation are compared which produces direct axis component of reference signal. The signals from d-q-o frame are again converted to a-b-c frame are compared with a filter current (I_{shabc}), which results in generation of reference compensation current, which is given as input to the hysteresis controller.

Figure 4 shows operating principle of hysteresis band controller is to produce triggering signals required for switching ON/OFF of IGBT's of shunt active filter. The objective of this controller is to control the compensation currents by forcing it to follow the reference ones. The switching strategies of the three-phase inverter will keep the currents into the hysteresis band. The real load currents are sensed and their non active components are compared with the reference compensation currents. The hysteresis comparator outputs signals are used to turn on the inverter power switches.

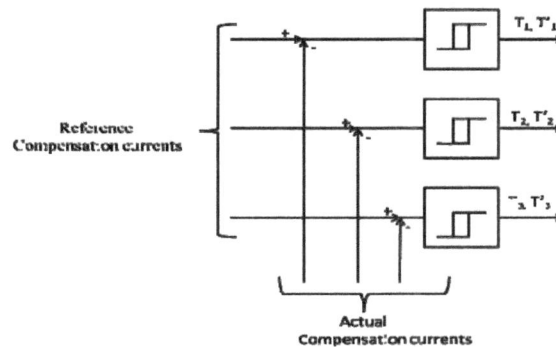

Figure 4. Operating principle of hysteresis band controller

4. RESULTS AND DISCUSSIONS
4.1. For Uncompensated System

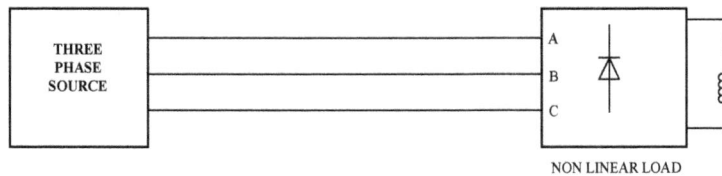

Figure 5. Uncompensated system

Figure 5 shows the simulation circuit for 3-phase 3-wire distribution system with a 3-phase voltage source connected to non linear load. Table 1 shows the various parameters of the considered system. Figure 6 shows Wave forms of source current and load current of uncompensated system. It can be observed from Figure 6 that instead of the actual sinusoidal waveform, a huge distortion in the source current can be observed. A delay can be observed in the output wave form, it is caused due to an inductor because an inductor opposes the sudden change in the current, though the supply wave form changes instantaneously it takes time for the inductor causing the delay in the wave form. Figure 7 shows the FFT analysis of source current. From the FFT analysis of the output waveform without filter shown in Figure 7, the %THD is about 29.71.

Table 1. System parameters

System parameters	Values Used
Source Impedance	L=0.01e-3 mH
Load	R=10Ω,L=30e-3 mH
Active filter	R=0.1 Ω,L=3e-3 mH

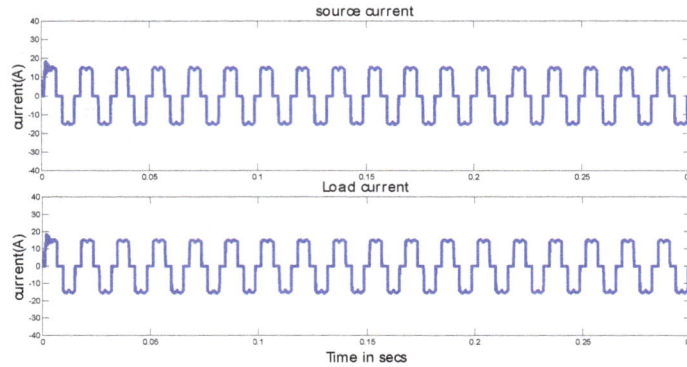

Figure 6. Wave forms of load current and source current of uncompensated system

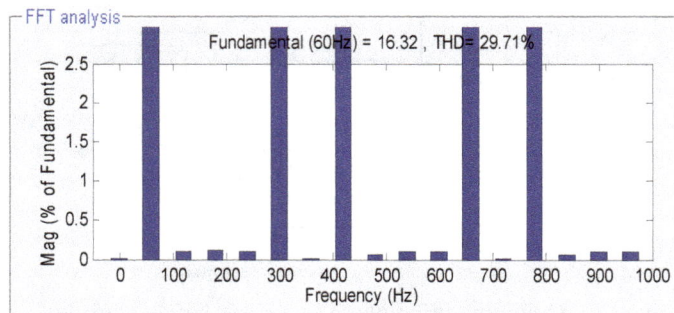

Figure 7. FFT analysis of source current

4.2. For Shunt Active Filter with ANN Controller

Figure 8 shows simulation circuit of Shunt Active Filter. Figure 9 shows the Simulation results of Shunt Active Filter with ANN Controller. From Figure 9, it can be observed that after Shunt Active Filter with ANN Controller runs, it reduces the much delay and waveform appears sinusoidally with fewer distortions when compared to uncompensated system and it also observed that the harmonics of the source current are eliminated by injecting the capacitor current which happens because of maintaining the capacitor voltage near to constant. Capacitor voltage takes 0.08sec to reach the steady state. Figure 10 shows FFT analysis of source current with ANN controller. From Figure 10, it can be seen that the current total harmonic distortion reduces to 2.27% from 29.71%.

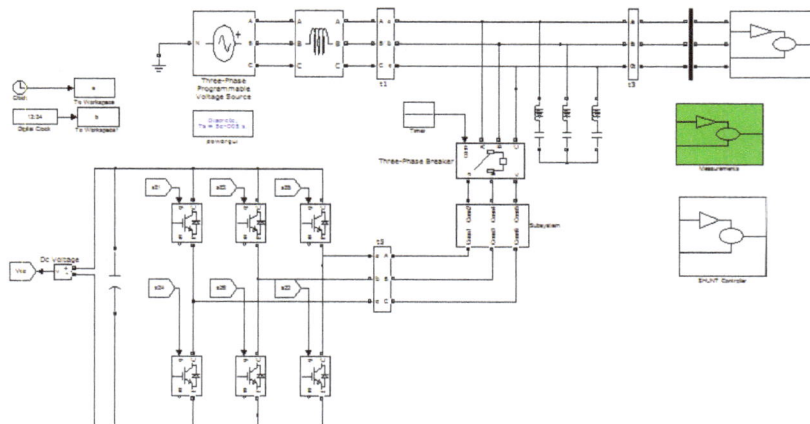

Figure 8. Simulation circuit of Shunt hybrid Active Filter

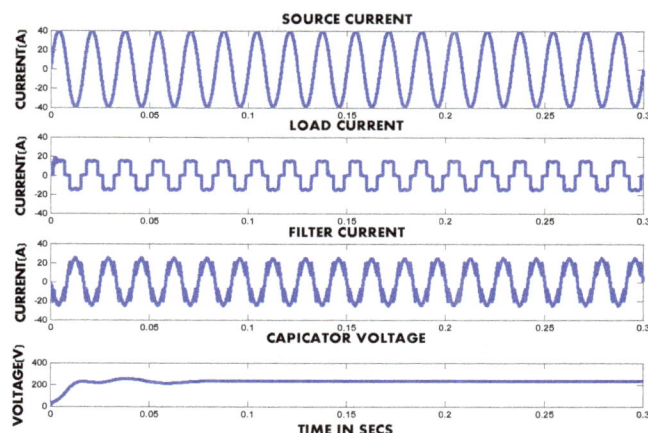

Figure 9. Simulation results of Shunt Active Filter with ANN Controller

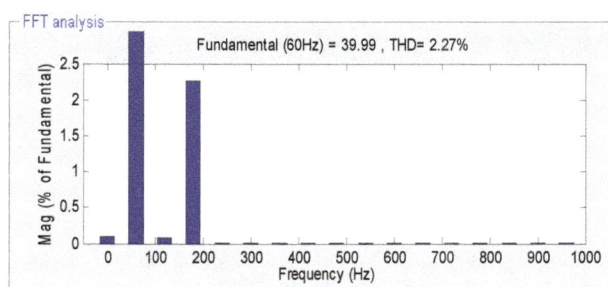

Figure 10. FFT analysis of source current with ANN controller

Table 2. Comparison of simulated results

| | SIMULATED RESULTS APF | |
	Uncompensated System	ANN
Settling Time (V_{DC}) in Sec.	--	0.08sec
%THD	29.71	2.27

5. CONCLUSION

In this paper, a detailed analysis of Shunt Active Power Filter with ANN controller has been proposed to mitigate harmonics of the three phase distribution system.

The obtained results show the simplicity and the effectiveness of the proposed intelligent controller under nonlinear load conditions. From the results, it can be observed that the current total harmonic distortion reduces better with ANN controlled active filter. The simulation and experimental results also show that the new control method is not only easy to be calculated and implemented, but also very effective in reducing harmonics

ACKNOWLEDGEMENTS

Following authors are highly supported and encouraged by the following institution by providing sufficient time and resources.

REFERENCES

[1] H Doğan, R Akkaya. *A Simple Control Scheme for Single-Phase Shunt Active Power Filter with Fuzzy Logic Based DC Bus Voltage Controller*. Proceedings of the International Multi Conference of Engineers and Computer Scientists. IMECS. Hong Kong. 2009; II

[2] Boren Ren lin, Richard G Hoft. Power electronics inverter with neural networks. *IEEE Technology update series.* 1997; 12(2).

[3] YM Chen, RM O'Connell. Active power line conditioner with a neural network control. *IEEE Trans. Ind. Appl.,* 1997; 33(4): 1131–1136.

[4] JR Vázquez, Patricio Salmeron, F Javieralcantra, Jaine Prieto. A new active power line conditioner for compensation in un unbalanced /distorted electrical power system. 14th PSCC, Sevilla. 2002.

[5] N Pecharanin. *An application of neural network to harmonic detection in active filter.* Proc. WCCI- ICNN. 1994; 6: 3756–3756.

[6] N Pechanranin, H Uitsui, M Sone. *Harmonic detection by using neural network.* Proc. IEEE Int. Conf. Neural Network., 1995, vol. 2, pp. 923–926.

[7] D Gao, Xiaoruisun. *A new method to generate current reference for active power filter.* Proc. IEEE CIEP, Alcapulco, Mexico. 2000: 99–103.

[8] Avik Bhattacharya, Chandan Chakraborty. A Shunt Active Power Filter With Enhanced Performance Using ANN-Based Predictive and Adaptive Controllers. *IEEE Transactions On Industrial Electronics.* 2011; 58(2).

[9] Mriul Jha, Satya Prakash Dubey. *Neuro-Fuzzy based controller for a Shunt Active Power Filter.* IEEE Conference, 2011.

[10] M Farahani. Damping of sub synchronous oscillations in power system using static synchronous series compensator. Published in IET Generation, Transmission & Distribution.

9

The Self Excited Induction Generator with Observation Magnetizing Characteristic in the Air Gap

Ridwan Gunawan*, Feri Yusivar*, Budiyanto Yan**

* Department of Electrical Engineering, University of Indonesia, Depok 16424, Indonesia

** Department of Electrical Engineering, University of Muhammadiyah, Jakarta 10510

ABSTRACT

Keyword:

dq0 transformation
Induction Machine
Magnetization
SEIG
State Space

This paper discusses The Self Excitated Induction Generator (SEIG) by approaching the induction machine, physically and mathematically which then transformed from three-phase frame abc to two-axis frame, direct-axis and quadratur-axis. Based on the reactive power demand of the induction machine, capacitor mounted on the stator of the induction machine then does the physical and mathematical approach of the system to obtain a space state model. Under known relationships, magnetization reactance and magnetizing current is not linear, so do mathematical approach to the magnetization reactance and magnetization current characteristic curve to obtain the magnetization reactance equation used in the calculation. Obtained state space model and the magnetic reactance equation is simulated by using Runge Kutta method of fourth order. The equations of reactance, is simulated by first using the polynomial equation and second using the exponent equation, and then to compare those result between the polynomial and exponent equations. The load voltage at d axis and q axis using the polynomial lags 640µs to the exponent equation. The polynomial voltage magnitude is less than 0.6068Volt from the exponent voltage magnitude.

Corresponding Author:

Ridwan Gunawan,
Department of Electrical Engineering,
University of Indonesia,
Depok 16424, Indonesia.
Email: ridwan@eng.ui.ac.id

1. INTRODUCTION

The length of the air-gap has a significant influence on the characteristics of an electric machine, the air-gap length has to be increased considerably from the value obtained for a standard electric motor. The efficiency of motor is highly dependent on the rotor eddy current losses. Air gap flux of induction motors contains rich harmonics. A flux monitoring scheme can give reliable and accurate information about electrical machine conditions. Any change in air gap, winding, voltage, and current can be reflected in the harmonic spectra [7]. A minimum air gap flux linkage is required for the self-excitation and stable operation of an isolated induction generator feeding an impedance load. The minimum air gap flux linkage requirement is the value at which the derivative of the magnetizing inductance with respect to the air gap flux linkage is zero. This minimum air gap flux linkage determines the minimum or maximum load impedance and minimum excitation capacitance requirements. This result is demonstrated using single-phase and three-phase induction generators [1]. Connection of induction generators to large power systems to supply electric power can also be achieved when the rotor speed of an induction generator is greater than the synchronous speed of the air-gap revolving field [2]. The magnetization curve of the induction motor was identified and compared with the one obtained by the no-load test. The method sensitivity to the load torque and the transient inductance has also been considered. A very good accuracy of the magnetization curve estimation has been also obtained at bigger load torques [3]. Two modes of operation can be employed for an induction generator. One is through self-excitation and other is through external-excitation. In first mode, the induction generator takes its excitation from VAR generating units, generally realized in the form of capacitor banks [4], [6].

The asynchronous machine is machine, what it has the rotating rotor and rotating stator flux's are different. The asynchronous machine is known also as the induction machine, what it does as a generator needed. One of specialty the induction generator from the synchronous generator can operated above synchronous speed, whice known as Self-Excitation. In this condition, the generator will use the energy, that it is generated from rotor rotation for to generate stator flux and rotor flux using reactive power. The reactive power is given local bank capacitor, that it conected to the stator. With suitable capacitors connected across the terminals and with rotor driven in either direction by a prime mover, voltage builds up across the terminals of the generator due to self excitation phenomenon leaving the generator operating under magnetic saturation at some stable point. Such generator is known as self-excited induction generator (SEIG) [4]. Using the simulation will be done the mathematical approach for hope to achieve a describe about all SIEG responses, in dq axis.

2. METHODOLOGY

The three phase induction generator has some equation. The equation flux average $\bar{\phi}$ is the flux as time function $\lambda(t)$ [11].

The equations stator voltage:

$$v_{as} = i_{as}r_s + \frac{d\lambda_{as}}{dt} \tag{1}$$

$$v_{bs} = i_{bs}r_s + \frac{d\lambda_{bs}}{dt} \tag{2}$$

$$v_{cs} = i_{cs}r_s + \frac{d\lambda_{cs}}{dt} \tag{3}$$

The equations rotor voltage:

$$v_{ar} = i_{ar}r_r + \frac{d\lambda_{ar}}{dt} \tag{4}$$

$$v_{br} = i_{br}r_r + \frac{d\lambda_{br}}{dt} \tag{5}$$

$$v_{cr} = i_{cr}r_r + \frac{d\lambda_{cr}}{dt} \tag{6}$$

The stator and rotor turns flux are written as below:

$$\begin{bmatrix} \lambda_s^{abc} \\ \lambda_r^{abc} \end{bmatrix} = \begin{bmatrix} L_{ss}^{abc} & L_{sr}^{abc} \\ L_{rs}^{abc} & L_{rr}^{abc} \end{bmatrix} \begin{bmatrix} i_s^{abc} \\ i_r^{abc} \end{bmatrix} \tag{7}$$

$$\lambda_s^{abc} = (\lambda_{as}, \lambda_{bs}, \lambda_{cs})^T \tag{8}$$

$$\lambda_r^{abc} = (\lambda_{ar}, \lambda_{br}, \lambda_{cr})^T \tag{9}$$

The stator and rotor current are written as below:

$$i_s^{abc} = (i_{as}, i_{bs}, i_{cs})^T \tag{10}$$

$$i_r^{abc} = (i_{ar}, i_{br}, i_{cr})^T \tag{11}$$

The Inductance stator to stator:

$$L_{ss}^{abc} = \begin{bmatrix} L_{ls} + L_{ss} & L_{sm} & L_{sm} \\ L_{sm} & L_{ls} + L_{ss} & L_{sm} \\ L_{sm} & L_{sm} & L_{ls} + L_{ss} \end{bmatrix} \tag{12}$$

The Inductance rotor to rotor:

$$L_{rr}^{abc} = \begin{bmatrix} L_{lr} + L_{rr} & L_{rm} & L_{rm} \\ L_{rm} & L_{lr} + L_{rr} & L_{rm} \\ L_{rm} & L_{rm} & L_{lr} + L_{rr} \end{bmatrix}$$

(13)

The Inductance stator to rotor and rotor to stator:

$$L_{sr}^{abc} = \left[L_{rs}^{abc}\right]^T$$

(14)

$$L_{sr} = \begin{bmatrix} \cos\theta_r & \cos\left(\theta_r + \frac{2\pi}{3}\right) & \cos\left(\theta_r - \frac{2\pi}{3}\right) \\ \cos\left(\theta_r - \frac{2\pi}{3}\right) & \cos\theta_r & \cos\left(\theta_r + \frac{2\pi}{3}\right) \\ \cos\left(\theta_r + \frac{2\pi}{3}\right) & \cos\left(\theta_r - \frac{2\pi}{3}\right) & \cos\theta_r \end{bmatrix}$$

(15)

Where:

stator self inductance   : $L_{ss} = N_s^2 P_g$ (1)

rotor self inductance   : $L_{rr} = N_r^2 P_g$ (2)

stator mutual inductance   : $L_{sm} = N_s^2 P_g \cos\left(2\pi/3\right)$ (18)

rotor mutual inductance   : $L_{rm} = N_r^2 P_g \cos\left(2\pi/3\right)$ (19)

stator to rotor peak mutual inductance $L_{sr} = N_s N_r P_g$ (20)
N_s : stator total turns lilitan stator
N_r : rotor total turns
P_g : air gap permeability

The equation transformation from stator and rotor in $qd0$ axis is obtained from the   Clark and Park transformation,

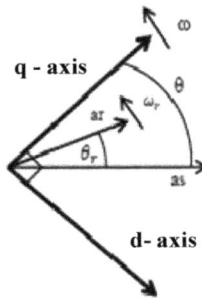

Figure 1. The "vector a-axis" at   stator and rotor and dq axis [10]

$$[fd \quad fq \quad fo]^T = \left[T_{dq0}(\theta)\right][fa \quad fb \quad fc]^T$$

(21)

The equation of stator and rotor position θ:

$$\theta(t) = \int_0^t \omega(t)dt + \theta_s(0)$$

(22)

$$\theta_r(t) = \int_0^t \omega_r(t)dt + \theta_r(0)$$

(23)

The matric transformation in dq0 axis, is shown as below:

$$[T_{dq0}(\theta)] = {}^2\!/_3 \begin{bmatrix} \cos\theta & \cos\left(\theta - \frac{2\pi}{3}\right) & \cos\left(\theta + \frac{2\pi}{3}\right) \\ \sin\theta & \sin\left(\theta - \frac{2\pi}{3}\right) & \sin\left(\theta + \frac{2\pi}{3}\right) \\ \frac{1}{2} & \frac{1}{2} & \frac{1}{2} \end{bmatrix} \tag{24}$$

And,

$$[T_{dq0}(\theta)]^{-1} = \begin{bmatrix} \cos\theta & \sin\theta & 1 \\ \cos(\theta - \frac{2\pi}{3}) & \sin(\theta - \frac{2\pi}{3}) & 1 \\ \cos(\theta + \frac{2\pi}{3}) & \sin(\theta + \frac{2\pi}{3}) & 1 \end{bmatrix} \tag{25}$$

The stator and rotor voltage in dq0 axis is shown:

$$v_{qs} = \dot{\lambda}_{qs} + \omega_s\lambda_{ds} + r_s i_{qs} \tag{26}$$

$$v_{ds} = \dot{\lambda}_{ds} - \omega_s\lambda_{qs} + r_s i_{ds} \tag{27}$$

$$v_{0s} = \dot{\lambda}_{0s} + r_s i_{0s} \tag{28}$$

$$v_{qr} = \dot{\lambda}_{qr} + (\omega_s - \omega_r)\lambda_{dr} + r_r i_{qr} \tag{29}$$

$$v_{dr} = \dot{\lambda}_{dr} - (\omega_s - \omega_r)\lambda_{qr} + r_r i_{dr} \tag{30}$$

$$v_{0r} = \dot{\lambda}_{0r} + r_r i_{0r} \tag{31}$$

The flux equation in dq0 axis:

$$\begin{bmatrix} \lambda_{qs} \\ \lambda_{ds} \\ \lambda_{0s} \\ \lambda_{qr} \\ \lambda_{dr} \\ \lambda_{0r} \end{bmatrix} = \begin{bmatrix} L_{ls} + L_m & 0 & 0 & L_m & 0 & 0 \\ 0 & L_{ls} + L_m & 0 & 0 & L_m & 0 \\ 0 & 0 & L_{ls} & 0 & 0 & 0 \\ L_m & 0 & 0 & L_{lr} + L_m & 0 & 0 \\ 0 & L_m & 0 & 0 & L_{lr} + L_m & 0 \\ 0 & 0 & 0 & 0 & 0 & L_{lr} \end{bmatrix} \begin{bmatrix} i_{qs} \\ i_{ds} \\ i_{0s} \\ i_{qr} \\ i_{dr} \\ i_{0r} \end{bmatrix} \tag{32}$$

The stator and rotor flux equations in dq0 axis:

$$\lambda_{qs} = L_s i_{qs} + L_m i_{qr} \tag{33}$$

$$\lambda_{ds} = L_s i_{ds} + L_m i_{dr} \tag{34}$$

$$\lambda_{qr} = L_r i_{qr} + L_m i_{qs} \tag{35}$$

$$\lambda_{dr} = L_r i_{dr} + L_m i_{ds} \tag{36}$$

Where:

$$L_s = (L_{ls} + L_m) \text{ and } L_r = (L_{lr} + L_m) \tag{37}$$

Analysis has been extended to identify effectiveness of the machine parameters to improve the operating performance of the generator. It is found that operating performance of the machine may be improved by proper design of stator and rotor parameters [4]. When an induction machine is driven by a prime mover, the residual magnetism in the rotor produces a small voltage that causes a capacitive current to flow. The resulting current provides feedback and further increases the voltage. It is eventually limited by the magnetic saturation in the rotor. Variable capacitance is required for self-excited induction generator [2]. The Self Excited Induction Generator (SEIG) using capacitors, is the induction generator as noload operation. This system is described as a three phase induction machine symetrically and conected to identic bank capacitor. The using model induction machine stationery, than to obtain equivalent circuit of the self excitated induction generator SEIG in d-axis, as Figure 2 as below [5]:

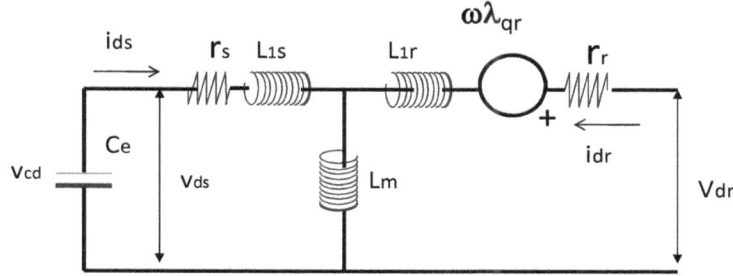

Figure 2. Stasionery circuit at d-axis with excited capacitor [5]

From the equivalent circuit as Figure 2, is obtained voltage equations in dq axis:

$$\begin{bmatrix} V_{ds} \\ V_{qs} \\ V_{dr} \\ V_{qr} \end{bmatrix} = \begin{bmatrix} r_s + L_s p + \frac{1}{pC_e} & 0 & L_m p & 0 \\ 0 & r_s + L_s p + \frac{1}{pC_e} & 0 & L_m p \\ L_m p & \omega L_m & r_r + L_r p & \omega L_r \\ -\omega L_m & L_m p & -\omega L_r & r_r + L_r p \end{bmatrix} \begin{bmatrix} i_{ds} \\ i_{qs} \\ i_{dr} \\ i_{qr} \end{bmatrix} \tag{38}$$

The three external elements that can change the voltage profile of SEIG are speed, terminal capacitance and the load impedance. By varying the elements, one at a time the performance characteristics of the squirrel-cage induction generator obtained. In most of SEIG applications, the rotational speed is rarely controllable. Therefore, the load seen by the generator or terminal capacitance has to be controlled [9]. The load RL series, is conected parallel with the bank capacitor.

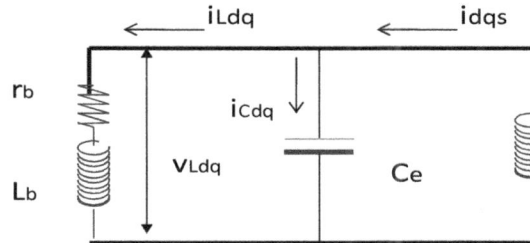

Figure 3. The SEIG RLC load [5]

$$i_{Cd} = C_e p \dot{V}_{Ld} = (r_b p C_e + L_b p^2 C_e) i_{Ld} \qquad and \qquad i_{Ld} = \frac{i_{ds}}{r_b p C_e + L_b p^2 C_e + 1} \tag{39}$$

Where:

$$i_{ds} = i_{Cd} + i_{Ld} , \quad and \ i_{ds} = (r_b p C_e + L_b p^2 C_e) i_{Ld} + i_{Ld}$$

$$V_{Ld} = (r_b + L_b p) \, i_{Ld} \quad or \quad V_{Ld} = \frac{r_b + p L_b}{r_b p C_e + L_b p^2 C_e + 1} \, i_{ds} \tag{40}$$

Using the the equivalent circuit for q-axis is obtained:

$$V_{Lq} = \frac{r_b + p L_b}{r_b p C_e + L_b p^2 C_e + 1} \, i_{qs} \tag{41}$$

The substitution voltage v_{cd}, v_{cq} and V_{Ld}, V_{Lq} to Equation (26)-(27) and (29)-(30), and then:

$$[v_{ds} \quad v_{qs} \quad v_{dr} \quad v_{qr}]^T = [Z] \begin{bmatrix} i_{ds} & i_{qs} & i_{dr} & i_{qr} \end{bmatrix}^T \tag{42}$$

$$Z = \begin{bmatrix} r_s + pL_s + \dfrac{r_b + pL_b}{r_b pC_e + L_b p^2 C_e + 1} & 0 & pL_m & 0 \\ 0 & r_s + pL_s + \dfrac{r_b + pL_b}{r_b pC_e + L_b p^2 C_e + 1} & 0 & pL_m \\ pL_m & \omega L_m & r_r + pL_r & \omega L_r \\ -\omega L_m & pL_m & \omega L_{lr} & r_r + pL_r \end{bmatrix} \tag{43}$$

The Equation (33) until (36) is written in the state space model, as below:

$$[\dot{x}] = [A][x] + [B][u] \tag{44}$$

Where:

$$[B][u] = K \begin{bmatrix} -L_r & 0 & L_m & 0 \\ 0 & -L_r & 0 & L_m \\ L_m & 0 & -L_s & 0 \\ 0 & L_m & 0 & -L_s \end{bmatrix} \begin{bmatrix} v_{ds} \\ v_{qs} \\ v_{dr} \\ v_{qr} \end{bmatrix} \tag{45}$$

$$[x] = [i_{ds} \quad i_{qs} \quad i_{dr} \quad i_{qr} \quad V_{Ld} \quad V_{Lq} \quad i_{Ld} \quad i_{Lq}]^T \tag{46}$$

$$K = 1/(L_m{}^2 - L_s . L_r) : \tag{47}$$

$$A = K \begin{bmatrix} A_{11} & A_{12} \\ A_{21} & A_{22} \end{bmatrix} \tag{48}$$

$$A_{11} = \begin{bmatrix} r_s L_r & -\omega L_m{}^2 & -r_r L_m & -\omega L_m L_r \\ \omega L_m{}^2 & r_s L_r & \omega L_m L_r & -r_r L_m \\ -r_s L_m & \omega L_m L_s & r_r L_s & \omega L_r L_s \\ -\omega L_m L_s & -r_s L_m & -\omega L_r L_s & r_r L_s \end{bmatrix} \quad A_{12} = \begin{bmatrix} L_r & 0 & 0 & 0 \\ 0 & L_r & 0 & 0 \\ -L_m & 0 & 0 & 0 \\ 0 & -L_m & 0 & 0 \end{bmatrix}$$

$$A_{21} = \begin{bmatrix} {}^1/_{C_e K} & 0 & 0 & 0 \\ 0 & {}^1/_{C_e K} & 0 & 0 \\ 0 & 0 & 0 & 0 \\ 0 & 0 & 0 & 0 \end{bmatrix} \quad A_{22} = \begin{bmatrix} 0 & 0 & {}^{-1}/_{C_e K} & 0 \\ 0 & 0 & 0 & {}^{-1}/_{C_e K} \\ {}^1/_{L_b K} & 0 & 0 & 0 \\ 0 & {}^1/_{L_b K} & 0 & 0 \end{bmatrix} \tag{49}$$

The reactance of inductance magnetizing Xm is determined using technical approach with the exponential equation as Equation (50):

$$X_m = \omega L_m = V_u / i_m = F. \left(K_1 e^{K_2 i_m{}^2} + K_3 \right) \tag{50}$$

Using the equations of reactance, is done an algorithm of simulation the self excited induction generator using Runge Kutta method. First using the polynomial equation and second using the exponent equation.

Simulation using Linear Time-Variying State Model is used discrete computation runge kutta method of fourth order, in the state space equation as below [8]:

$$\dot{x}(t) = f(x, t) \tag{51}$$

$$x[(n - 1)T] = x_T(n - 1) \text{ and } x_T(n) \cong x(nT) \tag{52}$$

The sampling time T is *step interval*. The state space counting programme is using the function $f(x, t)$ for determine $\dot{x}(t) = f(x, t)$ along x and t. And then is determine every step as below:

For $x_T(n + 1) \cong x((n + 1)T)$ (53)

$$f(x, t) = [A(x(t))][x(t)] + [B(x(t))][u(t)]$$ (54)

And,

$$[A(nT)] = [A_T(n)] = [A_T(x_T(n))]$$
$$[B(nT)] = [B_T(n)] = [B_T(x_T(n))]$$ (55)

$$[A((n + 1)T)] = [A_T(n + 1)] = [A_T(x_T(n + 1))]$$
$$[B((n + 1)T)] = [B_T(n + 1)] = [B_T(x_T(n + 1))]$$ (56)

$$g_o \equiv f[(x_T(n))]$$
$$g_1 \equiv f\left[(x_T(n)) + g_o\left(\frac{T}{2}\right)\right]$$
$$g_2 \equiv f\left[(x_T(n)) + g_1\left(\frac{T}{2}\right)\right]$$
$$g_3 \equiv f[(x_T(n) + g_2 T)]$$
$$g_4 \equiv (g_o + 2g_1 + 2g_2 + g_3)/6$$ (57)

Renew the state equation and time:

$$x_T(n + 1) = x_T(n) + g_4 T$$ (58)

$$[A_T(n + 1)] = [A_T(x_T(n + 1))]$$
$$[B_T(n + 1)] = [B_T(x_T(n + 1))]$$ (59)

$$n = (n + 1) \quad \text{and} \quad t = (n)T$$ (60)

$$x((n + 1)T) = f(x_T(n), nT)$$ (61)

$$f(x_T(n), nT) = [A(x_T(n))][x_T(n)] + [B(x_T(n))][u(n)]$$ (62)

3. RESULTS AND ANALYSIS

The data of the self excited induction generator SEIG, three phase 380 volt, 50 hertz, 7.5 kW, and 4 poles [5].

Tabel 1. Data of Self Exitated Induction Generator [5]

	magnitude	unit		magnitude	unit
r_s	1	Ohm	Ce	180	µF
L_s	1	mH	r_b	180	Ohm
r_r	0.77	Ohm	L_b	20	mH
Lr	1	mH	J	0.23	Kgm²

3.1 Simulation using the Polynomial Equation

In this simulation is used the magnetizing inductance equation [5]:

$$L_m = 0.1407 + 0.0014i_m - 0.0012i_m{}^2 + 0.00005i_m{}^3$$ (63)

Using the parameter in Table 1, is done some of simulations using Equation (53) and sampling time 10^{-4} second. The load voltage response in dq axis is shown as Figure 4 as below:

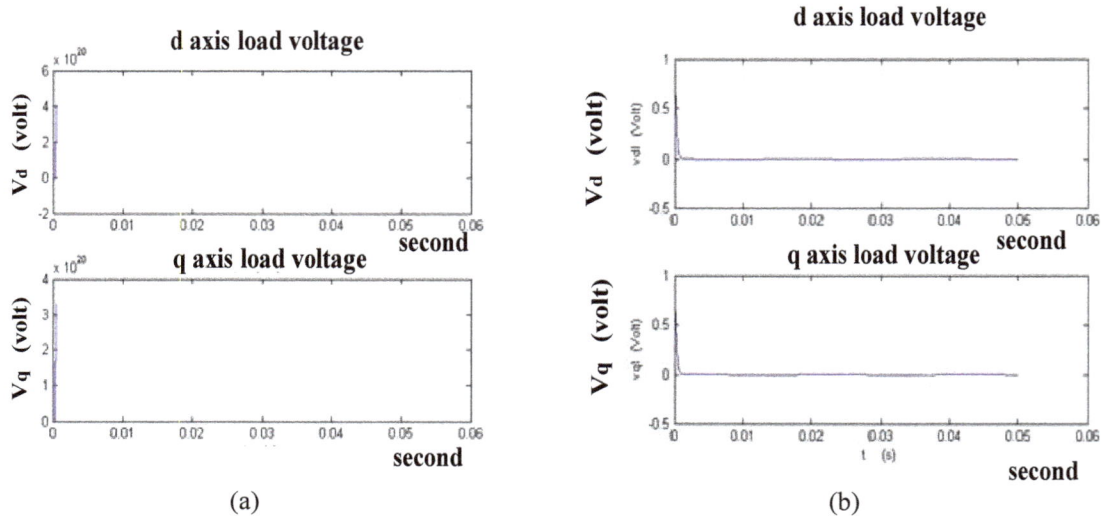

Figure 4. Load voltage using (a) sampling time 10^{-4} second, (b) sampling time 10^{-5} second

In Figure 4(a) the load voltage, that it the simulation does not give good response and not occure of the excitation because the sampling time very high. The second simulation is used the sampling time reduce to become 10^{-5} second, and the result is shown in Figure 4(b). The accuracy choice of sampling time gives a best response.

      After the excitation succeed, then using sampling time 10^{-5} second, is obtained the load voltage as Figure 5 as below:

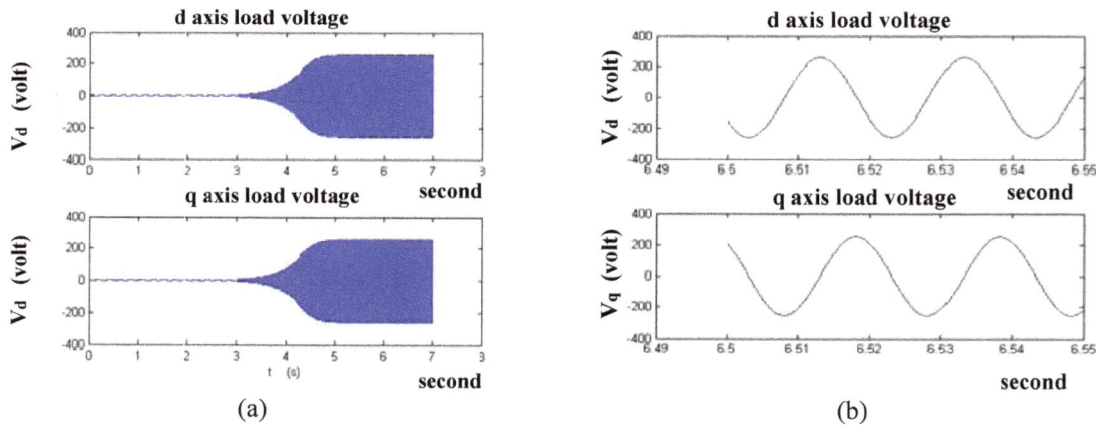

Figure 5. Load voltage at d axis and q axis (a) until 7 second. (b) time from 6.50 until 6.55 second

      The load voltage rises begin at time is 3 second until 5 second and after that its is constant at voltage 240 volt, and the form of wave is pure the sine form with the frequency is 50 cycles per second, as the conclusion this method using polynomial equation gives the accuracy response. The magnetizing curve from polynomial equation is shown in Figure 6, as below:

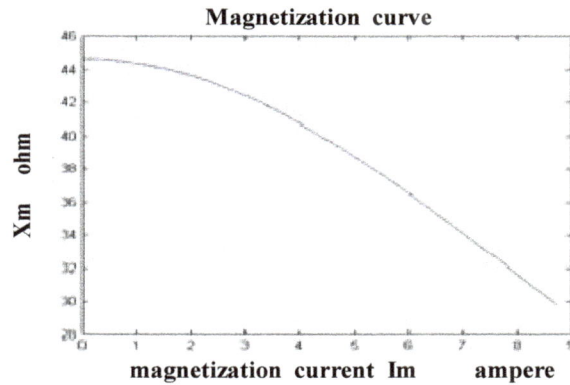

Figure 6. The magnetizing curve from polynomial equation

3.2. Simulation using the Exponent Equation

Base on using polinomial equation that it is iterrated by magnetizing current i_m in interval 0.01 ampere and then is determined the exponent equation using the programme "*constant Kij determine*", so that is obtained the curve as Figure 7, as below:

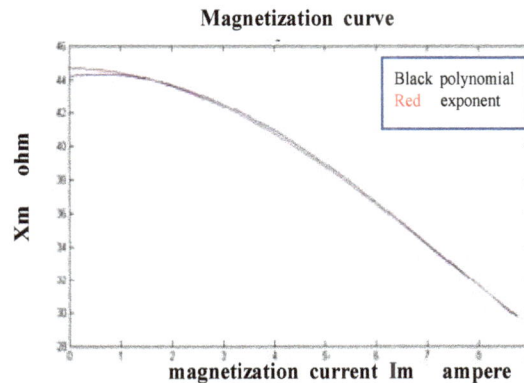

Figure 7. The polynomial equation using the exponent equation approach

Using the simulation and matlab programme is determined the constant K_1 = 0.1027 Ohm.second/radian , K_2 = -0.0081 1/ampere^2 and K_3= 0.0395 Ohm.second/radian. The magnetizing inductance curve L_m is shown in Figure 7 has exponent equation as Equation (64):

$$L_m = [0.1027 * (e^{-0.0081*i_m*i_m})] + 0.0395 \qquad (64)$$

Base on Figure 6, the magnetizing current starts from null ampere until 9 ampere. The constant value K_1, K_2 and K_3 is chosen, cause has a most precise value, that it nearst the polynomial equation until 9 ampere is shown as Figure 7. Using the data parameter motor and the exponent equation is done simulation , use the sampling time 10^{-4} second and the load voltage curve is shown as Figure 8(a). And then using the sampling time 10^{-5} second and do the simulation, is obtained the load voltage achieve the nominal voltage 240 volt, is shown as Figure 8(b):

The exponent equation of sinusoid load voltage at one periode is 20 milisecond, it means the frequency of sine wave is 50 cycles per second, and the peak load voltage achieves the nominal voltage 240 volt is shown in Figure 9. The result of simulation using the exponent equation is determine the magnetizing curve is shown as Figure 10.

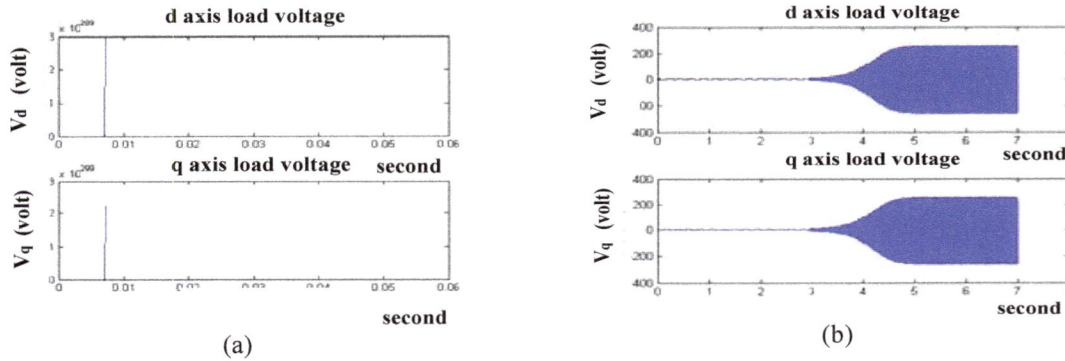

Figure 8. Load voltage using (a) sampling time 10^{-4} second, (b) sampling time 10^{-5} second

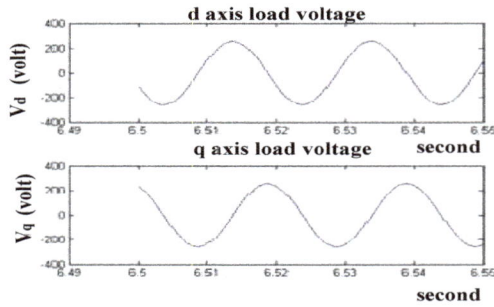

Figure 9. The load voltage at 6.50 until 6.55 second   Figure 10. The magnetizing curve

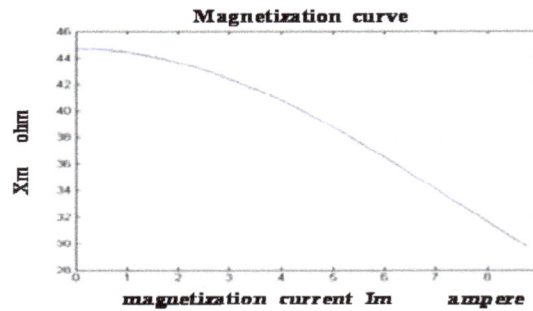

3.3. Comparison Results Between The Polynomial And The Exponent Equations

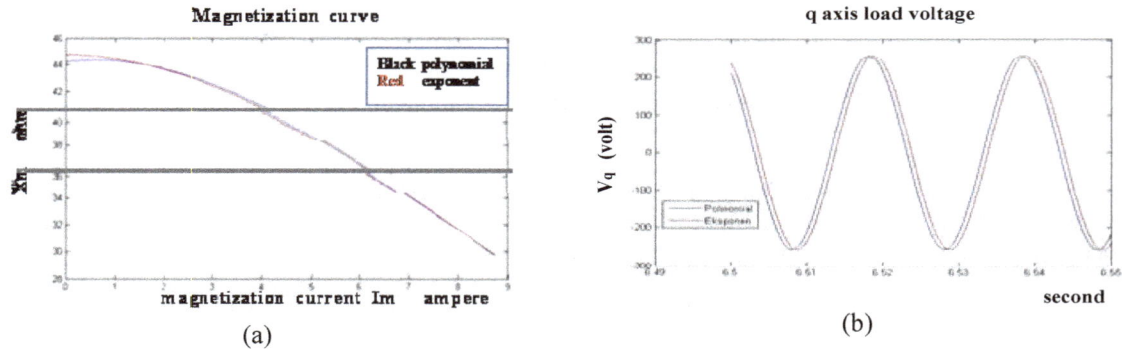

Figure 11. Comparison results using the polynomial and the exponent equation (a) the magnetism reactance Xm vs the magetizing current Im, (b) The loads voltage at q axis using the polynomial and the exponent equation

For to observe the difference between these results, is done to compare the data of it, that it are the magnetizing reactance Xm as function of and the magnetizing current i_m is shown as Figure 11(a). The second, to observe the difference results between the loads voltage using of both the equations. The load voltage at q-axis using the polynomial laggs 640 μs to the exponent equation and the polynomial voltage magnitude is less than 0.6068 volt from the exponent voltage magnitude. is shown as Figure 11(b).

4. CONCLUSION

The results have been determined for SEIG with using the iterration with sampling time , so much the smaller of sampling time, that the error value becomes very small. The accuracy choice of sampling time gives a best response. The phase to neutral output SEIG voltage for both equations achieves the nominal voltage and the form of waves are a sine wave. The comparison between the load voltages the polynomial and the exponent, the load voltage at using the polynomial laggs 640 μs to the exponent equation. The polynomial voltage magnitude is less than 0.6068 volt from the exponent voltage magnitude. The accuracy and exactness of magnetizing inductance equation L_m is very important, because these equations influence in determine of the equation in simulation depend on magnetizing current i_m. The accuracy of the magnetizing inductance L_m gives the output terminal voltage of SEIG is precision. The optimal distance of the air gap between stator and rotor, will be obtain the optimal magnetizing inductance. For the future this research can be expanded about noise in the air gap using the Wavelet Transform.

REFFERENCES

[1]   Ojo O. *Minimum Air Gap Flux Linkage Requirement for Self-Exitation in Stand Alone Induction Motor.* Energy Conversion, IEEE Transactions on. 2002; 10(3): 484 – 492.

[2]   VmSankardoss, SP Sabberwal, K Rajambal. *Experimental Design of Capasitance Require For Self-Excited Induction Generatos.* JATIT 15th. 2012: 45(1)

[3]   Božo Terzi´c, Marin Despalatovi´c, Alojz Slutej. *Magnetization Curve Identification of Vector-Controlled Induction Motor at Low-Load Conditions.* AUTOMATIKA: časopis za, 2012 - hrcak.srce.hr Applications. 2001; 37(6): 1801-1806.

[4]   Shelly Vadhera, KS Sandhu. *Estimation of saturation in grid connected induction generator.* International Journal on Emerging Technologies. 2010; (1): 31-36. ISSN : 0975-8364

[5]   K Trinadha, A Kumar, KS Sandu. *Study of Wind Turbine based SEIG under.* International Journal of Electrical and Computer Engineering (IJECE). 2012; 2.

[6]   Shelly Vadhera, KS Sandhu. *Estimation of saturation in grid connected induction generator.* International Journal on Emerging Technologies. 2010.

[7]   J Pyrhonen, M Valtonen, Aparviainen. Influence of the air-gap length to the performance of an axial-flux induction motor. Published in: Electrical Machines. ICEM. 18th International Conference. 2008.

[8]   M Godoy Simoes, Felix A Farret. *Renewable Energy Systems , Design and Analysis with Induction Generators.* CRC Press. 2004.

[9]   Subramanian Kulandhaivelu, KK RAYet. *Voltage Regulation of 3-Ø Self Excited Asynchonous Generator.* International Journal of Engineering Science and Technology. 2010; 2(12): 7215.

[10] Chee-Mun Ong. *Dynamic Simulation of Electrical Machinery.* New Jersey: Prentice Hall. 1998.

[11] P Albertos, A Sala, "*Multivariable Control Systems : An Engineering Approach.* Springer-Verlag London Limited, Valencia. 2004: 70.

Design and Experimental Verification of Linear Switched Reluctance Motor with Skewed Poles

N. C. Lenin*, R. Arumugam**
*VIT University, Chennai-600 127, Tamilnadu, India
**SSN College of Engineering, Chennai -603 110, Tamilnadu, India

ABSTRACT

Keyword:

FFT
Finite element analysis
Force ripple
Switched reluctance motor
Skewed pole

This paper presents the realization and design of a linear switched reluctance motor (LSRM) with a new stator structure. One of the setbacks in the LSRM family is the presence of high force ripple leads to vibration and acoustic noise. The proposed structure provides a smooth force profile with reduced force ripple. Finite element analysis (FEA) is used to predict the force and other relevant parameters. A frequency spectrum analysis of the force profile using the fast Fourier transform (FFT) is presented. The FEA and experimental results of this paper prove that LSRMs are one of the strong candidates for linear propulsion drives.

Corresponding Author:

N. C. Lenin,
Department of Electrical and Electronics Engineering,
Associate Professor, VIT University – Chennai Campus,
Chennai-600 127, Tamilnadu, India
Email: lenin.nc@vit.ac.in

1. INTRODUCTION

Linear switched reluctance motors (LSRMs) with different machine configurations have been explored past in the literature [1]–[12]. They are an attractive alternative to linear induction and linear synchronous machines due to lack of windings on either the stator or translator structure. However LSRM has some disadvantages such as high force ripple, vibration, and acoustic noise because of doubly salient structure. Moreover power electronic converters are required for their continuous operation. Efforts to reduce or eliminate the torque ripple of the rotary switched reluctance motors (SRMs) are presented in literature [13]-[17]. Multi phase excitation to reduce the force ripple in the LSRM has been explained in [18]. However the previous method considerably increases the copper losses. LSRM with pole shoes and inter poles are presented in [19]-[20]. In this paper a new stator structure [21] is proposed to reduce the force ripple.

Most of the limitations of analytical techniques can be overcome by using the numerical methods such as finite element analysis (FEA). These tools provide accurate results but require significant computational effort and numerical procedures [22]-[24]. The FEA tools are used in this study to predict the force and inductance profile.

When the frequency of the exciting force is close or equal to any of the natural frequencies of the machine, then resonance occur, which results in dangerous deformations and vibrations and a substantial increase in noise [25]. FFT steps to analyze ripple in the force profile of aLSRM is presented. This methodology is comparatively simpler than the most widely used finite-element vibration analysis procedure for mode frequency identification.

The organization of the paper is as follows: Section 2 and 3 presents new stator geometry for LSRMs that improves the force profile. In the new geometry, poles are skewed. Section 4 presents FEA results for conventional and proposed structures. Frequency spectrum analysis of force profile using the fast Fourier transforms (FFT) is highlighted in section 5. Experimental results from the prototype machine and their correlation with FEA results are presented in Section 6. Conclusions are summarized in Section 7.

2. LSRM TOPOLOGY

Figure 1 shows the two dimensional (2D) cross sectional view for the conventional machine structure of a three phase LSRM. The LSRM has an active translator and a passive stator. It consists of six translator poles and 120 stator poles. Figure 2 shows the stator pole alone for the conventional structure whereas, Figure 3 shows the stator pole for the proposed structure used for this study. The poles are skewed by an angle 1 degree to 10 degrees in steps of 1 degree for the purpose of optimization. Table 1 shows the physical dimensions of the LSRM prototype.

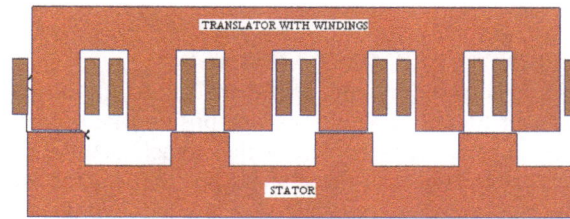

Figure 1. 2D cross sectional model of conventional LSRM

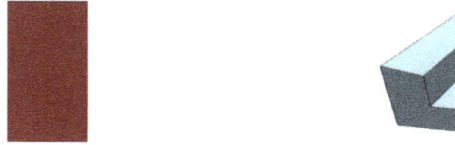

Figure 2. Conventional stator pole   Figure 3. Proposed stator pole

Table 1. Specifications of Prototype LSRM

Translator pole width=20mm	Translator slot width=20mm	Translator pole height=27mm
Translator back iron thickness=20 mm	No. of turns/phase = 86	Translator stack length=30 mm
Stator pole width=20mm	Stator slot width=26mm	Stator pole height=15 mm
Stator back iron thickness=20mm	Air gap length=2mm	Stator stack length=30 mm
Rated voltage=36 V	Rated current=4amps	Velocity=1m/s
Stator pole skewed angle = 1 to 10 degrees	Stator pole skew angle = 1 to 10 degrees	Maximum force=3.21N

3. INTRODUCTION TO FORCE RIPPLE IN LINEAR SWITCHED RELUCTANCE MOTOR

One of the inherent problems in LSRM is the force ripple due to switched nature of the force production. Force ripple may be determined from the variations in the output force. In order to predict the amount of force ripple, static force characteristics should be considered. The force dip is the distance between the peak value and the common point of overlap in the force angle characteristics of two consecutive LSRM phases as illustrated in Figure 4. Assuming that the maximum value of the static force F_{max} (peak static force) and the minimum value that occurs at the intersection point of two consecutive phases as F_{min}, the percentage force ripple may be defined as:

$$\%\text{Force Ripple} = \frac{F_{max} - F_{min}}{F_{avg}} \times 100 \tag{1}$$

The force dip is an indirect indicator of force ripple in the machine; the lesser the value of the force dip, the lesser will be the force ripple. The force dip of both the machines has been computed by FEA.

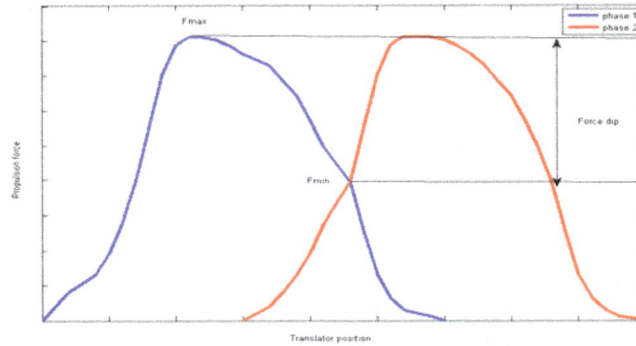

Figure 4. Force vs Translator position showing force dip

4. TWO – DIMENSIONAL FINITE ELEMENT ANALYSIS

Three asymmetric bridge metal oxide semiconductor field effect transistor (MOSFET) inverters are used to drive the LSRM shown in Figure 5. The translator position with respect to the stator position is sensed by three highly sensitive optical sensors. The active translator of the LSRM is moved from the unaligned position with respect to the stator to the aligned position for the excitation current of 4 amps. Therefore, static force and inductance profiles are obtained as a function of position and current.

Figure 6 shows the flux distribution taken from FEA for the conventional machine at the aligned position. The force and inductance profiles for the conventional and proposed LSRMs are depicted in Figure 7-10 respectively.

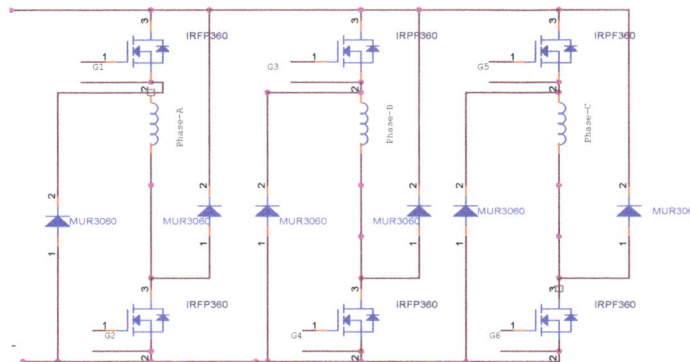

Figure 5. Three phase power converter for LSRM

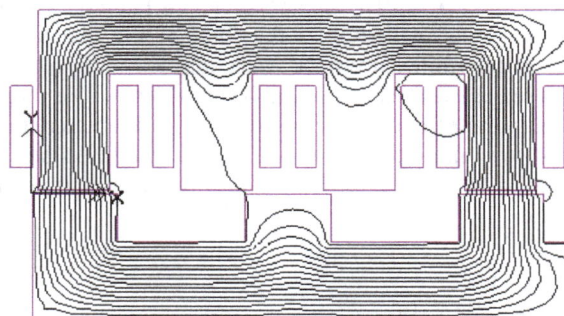

Figure 6. Flux distribution of conventional LSRM at the aligned position

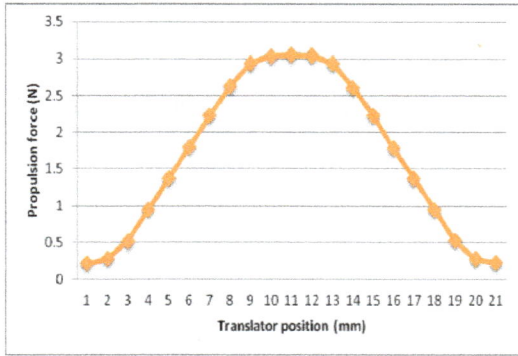

Figure 7. Propulsion force for base motor

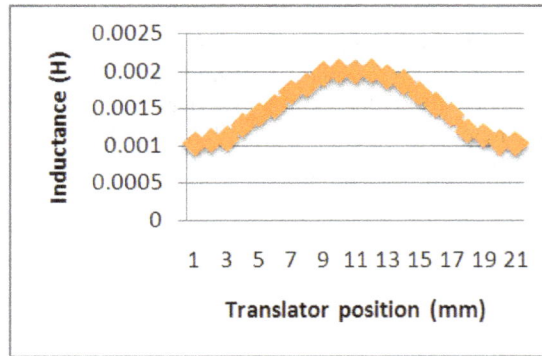

Figure 8. Inductance Profile for base motor

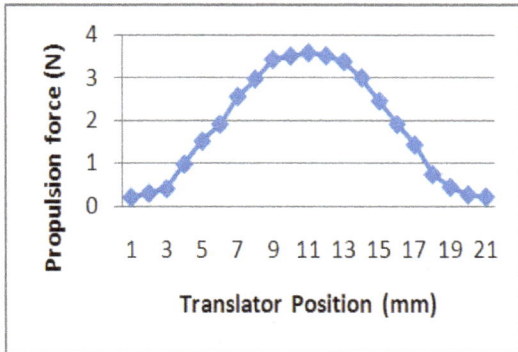

Figure 9. Propulsion force for proposed motor

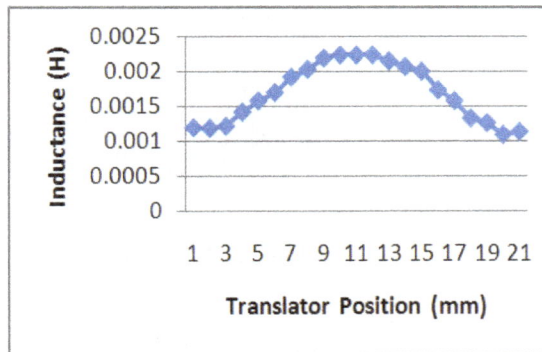

Figure 10. Inductance Profile for proposed motor

The force in a given direction is obtained by differentiating the magnetic co-energy of the system with respect to a virtual displacement of the translator. Based on this approach, the propulsion force with respect to various translator positions is calculated. The peak force obtained is 3.05N and the force ripple is 44.85% for the conventional LSRM whereas the peak force obtained in proposed LSRM is 3.32 N with the force ripple of 32.79%. The entire comparisons of the two structures are tabulated in Table 2.

Table 2. Summary of Comparison of the Two Structures

Type	Peak propulsion force (N)	Minimum propulsion force (N)	Average propulsion force (N)	Force ripple (%)	Inductance (H) Aligned	Unaligned
Conventional Stator	3.05	1.83	2.72	44.85	0.0020	0.001
Stator with skewed pole(6 degrees)	3.52	2.5	3.11	32.79	0.0023	0.0012

5. FAST FOURIER TRANSFORM APPLICATION TO LSRM

From the results of 2-D finite-element field analysis performed earlier, force (N) versus translator position (mm) will be known (Figure 6-10). A program is written in MATLAB environment which contains a sequence of instructions to store the force parameter array of the three phases. FFT is applied to the net force profile after the elimination of dc offset [26]. Since FFT transforms the available data in time domain into frequency domain, the available force versus translator position profile must be converted to force versus time profile. In MATLAB, the command fft(x,p), where 'x' is the force array and 'p' is 512, denoting 512 point fft, will solve the Equation (2) to produce a complex discrete fourier transform (DFT) of force. The absolute value of the obtained complex DFT will form the magnitude axis.

$$f(t) = \frac{1}{2\pi} \int_{\frac{-T}{2}}^{\frac{T}{2}} F(j\omega_0) \, e^{j\omega_0 t} dt \qquad (2)$$

The magnitude plot is obtained by plotting the magnitude versus frequency. Figure 11 shows the results of the frequency spectrum analysis for the conventional structure of the stator. The frequency corresponding to the decibel (dB) peaks can be identified from the plot. Table 3 shows the dominant frequencies in hertz and its amplitude in dB.

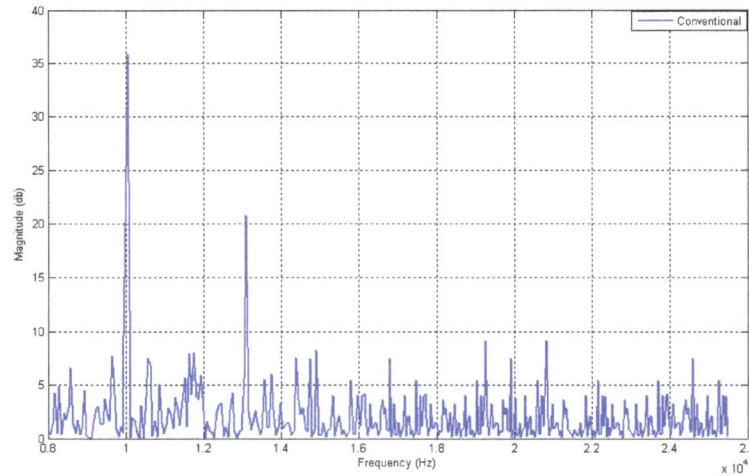

Figure 11. FFT output: dB versus frequency, for the LSRM

Table 3. Dominant Ripple Frequencies and its Amplitude for the Stator

Predominant ripple frequencies (Hz)	Amplitude (dB)
10,050	36
13,100	21
14,910	8
19,260	9.2

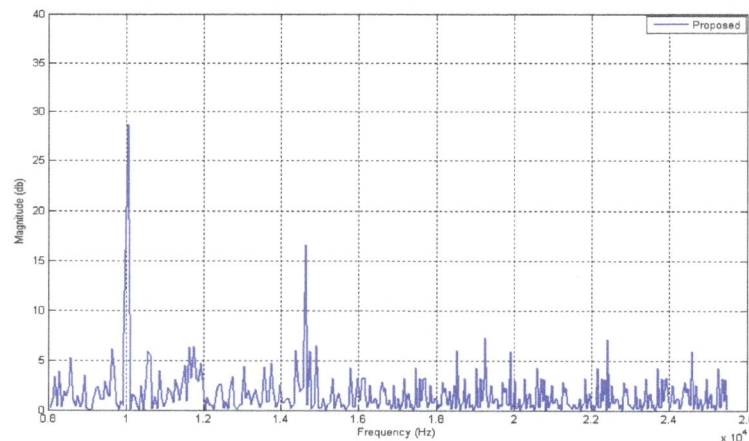

Figure 12. FFT output: dB versus frequency, for the LSRM with skewed poles

The result of FFT for the LSRM with skewed poles is depicted in Figure 12. It is observed that the magnitude dB peaks occurs at the same frequencies in both cases. However the magnitude of the dB peaks is reduced by a considerable margin, encouraging the design case of skewed stator poles (Table 4).

Table 4. Dominant Ripple Frequencies and its Amplitude for Stator with Skewed Poles

Predominant ripple frequencies (Hz)	Amplitude (dB)
10,050	29
14,620	17
19,260	7.4
22,420	7

6. EXPERIMENTAL RESULTS

Figure 13 shows the experimental setup for the prototype LSRM used as a material carrying vehicle in the laboratory. The experimental road is 0.5 m long and translator weight is 2.7kg. It should be noted that the present setup is intended for development purposes only.

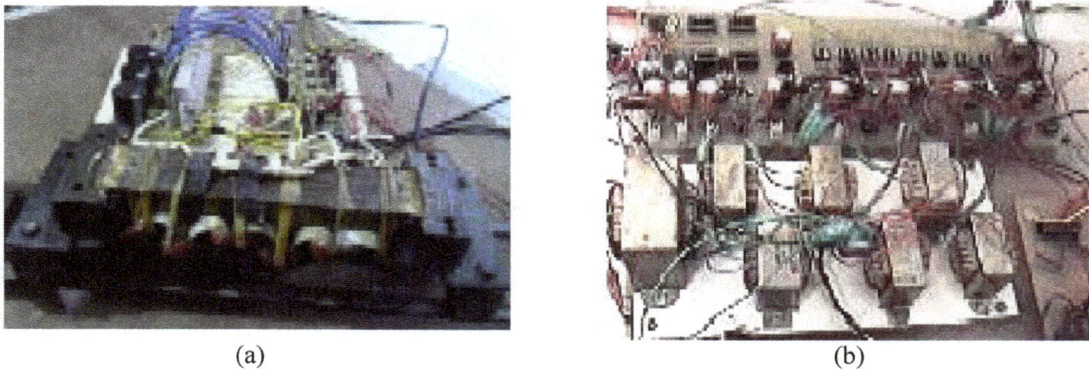

(a)   (b)

Figure 13. Experimental setup of (a) LSRM and converter (b) Driver circuit

The inductance for the different positions at rated current is measured by locking the translator at each position. A constant current is applied to a phase and is turned off and the falling current profile is computed. The time constant is measured from the profile and hence the inductance is calculated. The measured values of inductance and propulsion force are plotted alongside the FEA results in Figure 14 and Figure 15 respectively. Figure 16 shows phase current and pulse waveforms of the LSRM.

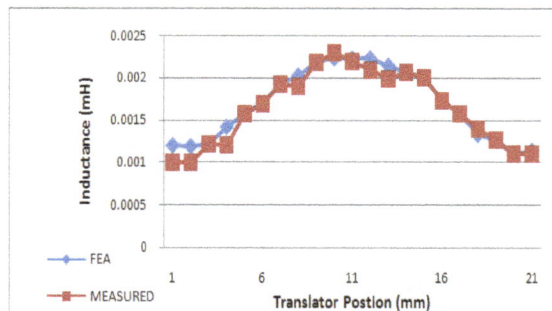

Figure 14. Comparison of FEA and measured inductance values at rated current

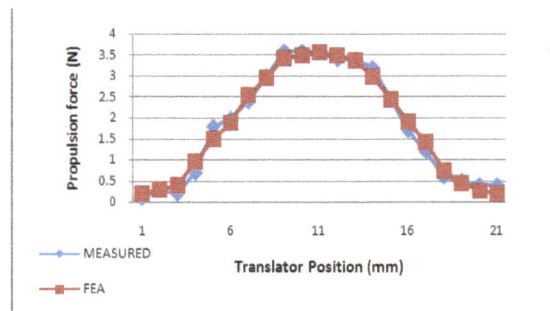

Figure 15. comparison of FEA and measured propulsion force at rated current

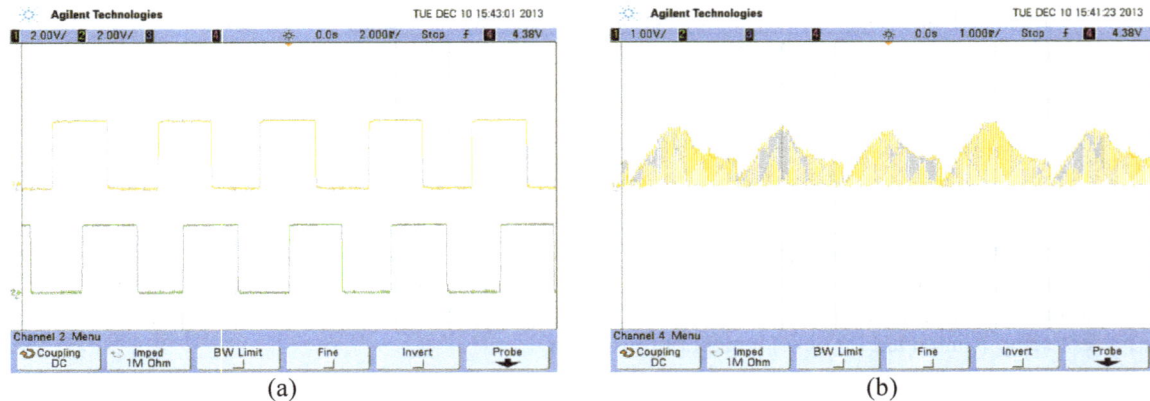

Figure 16. Experimental waveforms (a) Pulses of LSRM (b) Phase current of LSRM

7. CONCLUSION

Modification of the stator geometry by the provision of skewed poles has been presented in this paper. A 2m long LSRM prototype has been constructed. Force and inductance profile has been obtained by using FEA. There is a good agreement between measurement results and FEA values of inductance profile of the motor. The proposed structure reduces the force ripple by approximately 27% compared to the conventional machine. FFT methodology is comparatively simpler than the most widely used finite-element vibration analysis procedure for mode frequency identification.

REFERENCES

[1] J Corda, E Skopljak, *Linear switched reluctance actuator*, IEEE Proceedings, 1987; 125(4): 535-539.
[2] D Matt, R Goyet, J Lucidarme, C Rioux, Longitudinal-field multi-air gap linear reluctance actuator, *Elect. Mach. Power Syst.*, 1987; 13(5): 299–313.
[3] K Takayama, Y Takasaki, A new type switched reluctance motor, *IEEE Transactions on Industry Applications*, 1988; 23(5): 71-78.
[4] Mimpei Morishita, A new maglev system for magnetically levitated carrier system, *IEEE Transactions on Vehicular Technology*, 1989; 38(4): 230-236.
[5] PM Cusack, GE Dawson, TR Eastham, *Design, control and operation of a linear switched reluctance motor*, Proc. Canadian Conf. Electrical and Computer Engineering, Quebec, PQ, Canada, 1991; 19.
[6] J Corda, E Skopljak, *Linear switched reluctance actuator*, Proc. Sixth Int. Conf. Electrical Machines and Drives, Oxford, U.K, 1993; 535–539.
[7] J Lucidarme, A Amouri, M Poloujadoff, Optimum design of longitudinal field variable reluctance motors-application to a high performance actuator, *IEEE Trans. Energy Conversion*, 1993; 8: 357–361.
[8] US Deshpande, JJ Cathey, E Richter, A high force density linear switched reluctance machine, *IEEE Transactions on Industry Applications*, 1995; 31(2): 345-52.
[9] CT Liu, YN Chen, *On the feasible polygon classifications of linear switched reluctance machines*, Conf. Rec. 1997 IEEE Int. Electric Machines and Drives Conf., Milwaukee, WI, 1997; 11.
[10] BS Lee, HK Bae, P Vijayraghavan, R Krishnan, Design of a linear switched reluctance machine, *IEEE Transactions on Industry Application*, 2000; 36: 1571-1580.
[11] Ferhat Daldaban, Nurettin Ustkoyuncu, A new double sided linear switched reluctance motor with low cost, *Energy Conversion and Management*, 2006; 47: 2983–2990.
[12] Hong Sun Lim, Ramu Krishnan, Ropeless Elevator with linear switched reluctance motor drive actuation systems, *IEEE Transactions on Industrial Electronics*, 2007; 54(4): 2209 - 2218.
[13] DS Schramm, BW Williams, TC Green, *Torque ripple reduction of switched reluctance motors by phase current optimal profiling*, Proc. IEEE PESC'92, 1992; 857–860.
[14] M. Moallem, C. M. Ong, and L. E. Unnewehr, Effect of rotor profiles on the torque of a switched reluctance motor, *IEEE Trans. on Ind. Applicat.*, 1992; 28(2): 364-369.
[15] Iqbal Hussain, and M. Ehsani, Torque Ripple Minimization in Switched Reluctance Motor Drives by PWM Current Control, *IEEE Trans., on Power Electronics*, 1996; 11(1): 83-88.
[16] R. Rabinovici, Torque ripple, vibrations, and acoustic noise in switched reluctance motors, *HAIT Journal of Science and Engineering B*, 2005; 2(5-6): 776-786.
[17] C Neagoe, A Foggia, R Krishnan, *Impact of pole tapering on the electromagnetic force of the switched reluctance motor*, Conf. Rec IEEE Electric Machines and Drives Conference, 1997: WA1/2.1- WA1/2.3.
[18] P Silvester, MVK Chari, Finite element solution of saturable magnetic field problems, *IEEE Trans. On Power Apparatus and Systems (PAS)*, 1970; 89: 1642-1651.

[19] NC Lenin, R Arumugam, Analysis and characterisation of longitudinal flux single sided linear switched reluctance machines, *Turkish Journal of Electrical Engineering and Computer Sciences*, 2012; 20(1): 1220-1227.

[20] NC Lenin, R Arumugam, Vibration analysis and control in linear switched reluctance motor, *Journal of Vibroengineering*, 2011; 13(4): 662-675.

[21] NC Lenin, R Elavarasan, R Arumugam, *Investigation of Linear Switched Reluctance Motor with Skewed Poles*, International Conference on Advances in Electrical Engineering, 2014.

[22] JF Lindsay, R Arumugam, R Krishnan, *Magnetic field analysis of a switched reluctance motor with multitooth per stator pole*, Proc. Inst Elect Eng, 1986: 347–353.

[23] NN Fulton, *The application of CAD to switched reluctance drives*, Electrical Machines Drives Conf, 1987: 275–279.

[24] AM Omekanda, C Broche, M Renglet, Calculation of the electromagnetic parameters of a switched reluctance motor using an improved FEM–BIEM application to different models for the torque calculation, *IEEE Transactions on Industry Applications*, 1997: 914–918.

[25] DE Cameron, Jeffrey H Lang, SD Umans, The origin and reduction of acoustic noise in doubly salient variable-reluctance motors, *IEEE Transactions on Industry Applications*, 1992; 28(6): 1250–1255.

[26] KN Srinivas, R Arumugam, *Spectrum analysis of torque ripple in a switched reluctance motor*, Proc. Nat. Conf. Elect, Chennai, India, 2004; 88–93.

11

Voltage Flicker Mitigation in Electric Arc Furnace using D-STATCOM

Deepthisree M, Ilango K, Kirthika Devi V S, Manjula G Nair
Departement of Electrical and Electronics Engineering, Amrita Vishwa Vidyapeetham, Amrita School of Engineering, Bengaluru, India

ABSTRACT

Keyword:

D-STATCOM-Distribution
Static Synchronous
Compensator
EAF-Electric arc furnace
Power Quality
THD-Total harmonic distortion
Voltage Flicker

The major power quality issue of voltage flicker has resulted as a serious concern for the customers and heavy power companies. Voltage flicker is an impression of unsteadiness of visual sensation induced by a light source whose luminance fluctuates with time. This phenomenon is experienced when an Electric Arc Furnace (EAF) as load is connected to the power system. Flexible AC transmission devices (FACTS) devices were gradually utilized for voltage flicker reduction. In this paper the FACTS device of Distribution Static Synchronous Compensator (D-STATCOM) is used to serve the purpose of mitigating voltage flickering caused by electric arc furnace load, which is efficiently controlled by Icosϕ control algorithm. The model of electric arc furnace is considered as a current source controlled by a non linear resistance, which had been simulated and performance was analyzed using MATLAB/SIMULINK Software.

Corresponding Author:

Ilango.K,
Departement of Electrical and Electronics Engineering,
Amrita Vishwa vidyapeetham University,Amrita School of Engineering,
Kasavanahalli, Carmelaram (Post), Bengaluru, India- 560035
Email: kilango2002@gmail.com

1. INTRODUCTION

The quality of power is a great concern for the heavy power companies and end users. There are many power quality issues such as sag, swell, harmonics, switching transients, interruptions, overvoltage, under voltage, voltage flicker, low power factor etc. The quality of power has an economic impact on consumers. Nonlinearity in the load causes harmonics which can cause drastic effects on the power supply and adversely affect the performance of other electrical equipments. Electric Arc Furnace is one such equipment, when connected to the grid system results in voltage fluctuations which lead to the power quality problem of voltage flicker, power factor redection, and harmonic distortion in the electric grid. These problems reduces the electrical equipment efficiency. The problems caused by electric arc furnace can be solved by using many techniques, here D-STATCOM has been selected for mitigating voltage flicker and reduction of total harmonic distortion. Traditionally, SVC (Static Var Compensators) were used to solve the problem of voltage flicker, reactive powercompensation and power factor correction but it had few disadvantages that it generates lower order harmonics and takes longer response time compared to D-STATCOM. The D-STATCOM is an IGBT-based voltage source inverter which delivers faster response, power factor improvement and harmonic reduction, voltage sag mitigation since it uses highly advanced power switches [1]. D-STATCOM benefits in reducing the losses by maintaining high power factor at the load end, which eliminates voltage sags, swells and transients. If a D-STATCOM is well designed it serves as an advantageous investment for the power companies and can bring productive gains for the company as well as customers.

The control algorithm decides the functionality of D-STATCOM, various control strategies have been reported in literatures. The conventional algorithms took more computational time and were found to be slow in response which made it a necessity to go for new and faster control strategy for D-STATCOM [2]-[3]. In this paper IcosΦ control algorithm has been used for D-STATCOM to mitigate voltage flicker and power factor correction. This control algorithm is a very simple method which involves less comptutional time targeting to limit the voltage flicker at the source side. The detailed algorithm has been reported in [4] and its application to renewable energy source has been reported in [5]. The few modifications have been introduced with IcosΦ control algorithm which has been used for the control of D-STATCOM.

The power system model with electric arc furnace is shown in the Figure 1. The electric arc furnace load is connected to power system introducing a flicker. The D-STATCOM provides reactive power compensation, thereby stabilizing the power system voltage input and reducing the flickering. The modelling of electric arc furnace is described in Section 2. The control algorithm is explained in section 3. The simulation results and conclusion has been presented in section 4 and section 5 respectively.

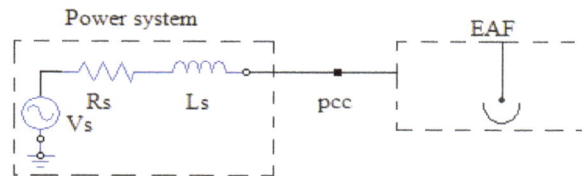

Figure 1. Power system model with Electric Arc Furnace

2. MODELLING OF ELECTRIC ARC FURNACE

The most prevalent way today to reprocess steel from scrap is electric arc furnace. A wide variety of scrap can be melted in the arc furnace to produce steel with the help of electrodes and current. The electrodes are moved in and out of the furnace and thereby arcing occurs which melts the ore. Since electric arc furnace is a highly nonlinear load, the modelling of electric arc furnace and its equivalent electrical model is a difficult task. Many literatures have proposed different models of electric arc furnace such as harmonic voltage source model, time domain model, frequency model etc [6].

The modeling of electric arc furnace mainly depends on the parameters such as voltage, current and length determined by the position of the electrodes which is moved in and out of the electric arc furnace. EAF has both AC and DC models. The operation of EAF can be explained with various stages of arc processes such as the melting and refining. In the first stage the electrode is pushed in and out of the furnace by means of an external control stabilizing the arc. During this processes a momentary short circuit is experienced at the secondary side of the arc furnace transformer. High fluctuation in current is experienced when the arcing occurs. These high fluctuations in the current cause a flicker in the voltage.

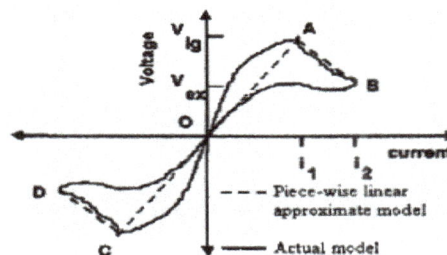

Figure 2. V-I characteristics of arc

The electric arc furnace is modelled as a current controlled time varying nonlinear resistance using MATLAB [7]-[8]. MATLAB Embedded program takes in two inputs, electric arc furnace current and the derivative of electric arc furnace current to the function giving a time varying non linear resistance as output. The MATLAB embedded program function block is interconnected to the grid by a current controlled source. In real time electric arc furnace is modelled as a dynamic model. The voltage current characteristic is shown in Figure 1 [7]. Sinusoidal variation is assumed for the time varying nonlinear resistance.

$$R_{arc}(t) = R_{arc}(1 + m_d \sin(wt)) \tag{1}$$

Where:

R_{arc} is the arc resistance

m_d is the modulation coeffiecient

w is the flicker frequency

$$R_s; for\ 0 \le |i| < i_g\ and\ \frac{d|i(t)|}{dt} > 0 \tag{2}$$

$$R_{arc} = \frac{V_d + (V_g - V_d)e^{\frac{-(|i| < i_g)}{t1}}}{|i|}; \tag{3}$$

$$for\ |i| \ge i_g\ and\ \frac{d|i(t)|}{dt} > 0$$

$$R_{arc} = \frac{V_t + (V_g - V_t)e^{\frac{-|i|}{t2}}}{(|i| + i_g)}; for\ \frac{d|i(t)|}{dt} < 0 \tag{4}$$

Where,

$$V_g = 1.15 * V_d \tag{5}$$

$$i_g = \frac{V_g}{R_s} \tag{6}$$

$$V_t = \frac{(I_{max} + i_g)}{I_{max}} * V_d \tag{7}$$

R_s : Slag resistance

V_g : Arc voltage

i_g : Arc current

t_1 : Time constant during melting process

t_2 : Time constant during arc extinction process

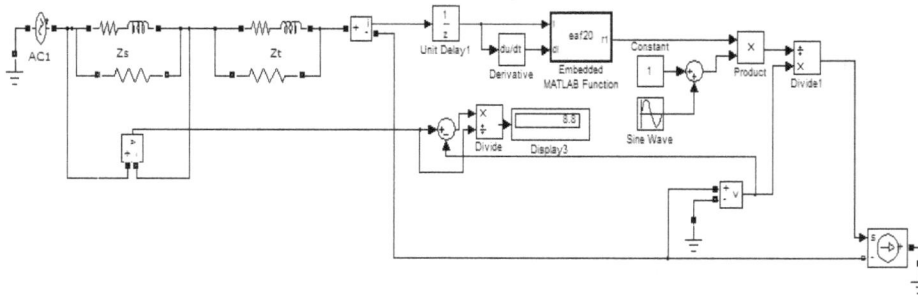

Figure 3. Simulink model of Single phase EAF model

3.   CONTROL ALGORITHM

The D-STATCOM is connected at the point of common coupling and it is driven by Icosϕ control algorithm. A simple method for achieving power factor correction and reduction of total harmonic distortion has prompted the proposal of Icosϕ control algorithm [4]. The Icosϕ is a less complicated and easy to implement algorithm. This algorithm is based on fundamental component of real part of load current .The current measured at the load side and voltage measured at the source side is given as input of controller. When load current passedd through a second order low pass filter obtains fundamental load current with a phase shift of 90^0 using bi-quad fileter. Unit amplitude source voltage is given to the 'detect fall negative block' which detects the zero crossing instant of the source voltage. Both the above mentioned current and voltage is moved as inputs to the sample and hold circuit.The magnitude of | Icosϕ| is received as output from the sample and hold circuit which is multiplied with the unit amplitude voltage to produce the desired source reference current at each phase. The desired source reference current is subtracted from the load current to obtain the reference compensation current. While generating compensation current, DC link voltage at the D-STATCOM side is balanced with help of pi controller which is added to the reference compensation current.

For a balanced source, the instantaneous voltages can be specified with a phase difference of 120^0. Equations for phase 'a' are explained below.

e_a =source voltage for phase a.

i_{la}= load current for phase a.

ϕ_a = phase angle of fundamental current in phase a.

$i_{la1}, i_{lb1}, i_{lc1}$= fundamental current amplitude for phase a,b,c.

U_a = unit amplitude of source voltage of phase a.

i_a= magnitude of desired current.

i_{da}=desired current

i_{ca} =compensation current

e_{dc}= dc link voltage at D-STATCOM terminal

e_{dcref}=dc reference voltage

$$e_a = E_m sin\omega t \tag{8}$$

$$|R_e(i_{la1})| = |i_{la}| * cos\phi_a \tag{9}$$

$$U_a = 1. sin\omega t \tag{10}$$

$$|i_a| = \frac{|R_e(i_{la1})|+|R_e(i_{lb1})|+R_e(i_{lc1})|}{3} \tag{11}$$

$$i_{da} = |i_a| * sin\omega t \tag{12}$$

$$i_{ca} = i_{la} - i_{da} \tag{13}$$

Error at the nth sampling instant

$$error(n) = e_{dcref} - e_{dc}(n) \tag{14}$$

Error is fed to PI controller to generate i_{caerr}.

$$i_{caerr}(n) = \{ i_{caerr}(n-1) + k_p (error(n) - error(n-1)) + k_i(error(n))\} \tag{15}$$

i_{comp} Current is passed to the hysteresis controller to generate pulses for D-STATCOM.

$$i_{comp} = i_{ca} + i_{caerr} \tag{16}$$

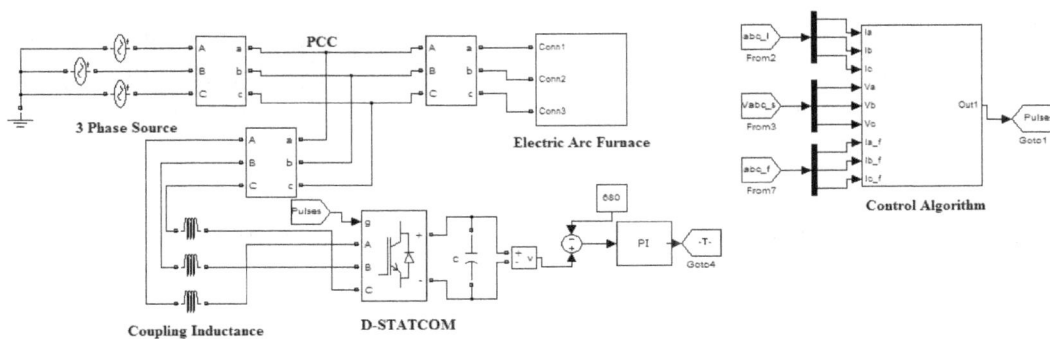

Figure 4. Simulink model of three phase system supplying Electric Arc Furnace load with D-STATCOM

The electric arc furnace was designed to 30 MW which is connected to the three phase electric grid.The D-STATCOM and its control algorthim were modeled and developed using MATLAB which was presented in Figure 4. The system was designed with simulation parameters shown in Table 1.

Table 1. Simulation Parameters

V_{source}	500V
f	50Hz
Zs	0.000528+j0.00468Ω
Zt	0.0003366+j0.00322Ω
Rs	0.05Ω
Vg	350.75V
Imax	100KA
t1	0.01sec
t2	0.01sec
DC link voltage	680V
DC link capacitor	500000µf
Coupling Inductance	8mH

4.   SIMULATION RESULTS

The described system was simulated and the results have been analyzed. The Figure 5 depicts the arc furnace load voltage waveform at source side. From Figure 5, it was clear that, flickering has been experienced when the electric arc furnace was connected at the load side. The percentage of flickering in the voltage without connecting D-STATCOM was found to be 8.8%.

The Figure 6 shows the arc resistance waveform of arc furnace. Figure 7 depicts the arc current wave- form which has the average magnitude of 60kA.

The Figure 8 shows the source voltage waveform when the D-STATCOM was connected to the network at the point of common coupling. From Figure 8, it is clear that the source voltage wave is a pure sinusodial without flickering effect.

The arc furnace load current THD analysis has been shown in Figure 9 which contains 20.91% of harmonics before compensation using D-STATCOM. From Figure 10, it can be observed that the THD analysis of arc furnace load current after compensation has reduced to 1.69%.

After compensation the source voltage and source current waveform has been depicted in Figure 11 which shows that the voltage and current were in-phase with each other. This was evident for power factor correction using D-STATCOM.

The arc furnace source voltage THD analysis has been carried out and shown in Figure 12, which contains harmonics of 33.02%. From Figure 13, it can be seen that the source voltage THD of arc furnace hafter compeas been reduced to 0.02% from 33.02%.

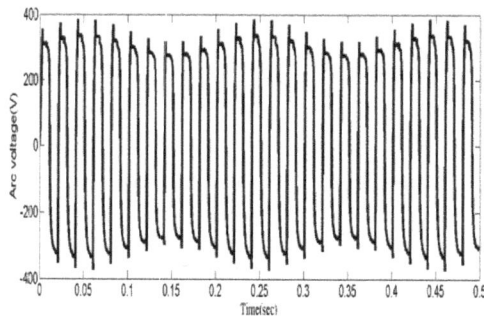

Figure 5. Arc voltage before compensation

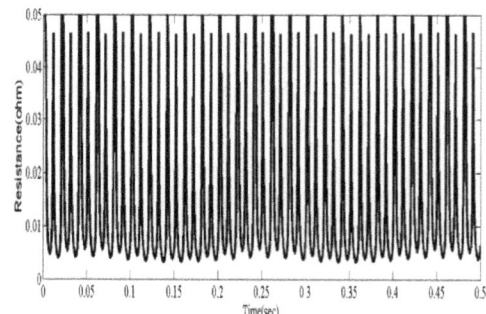

Figure 6. Arc resistance

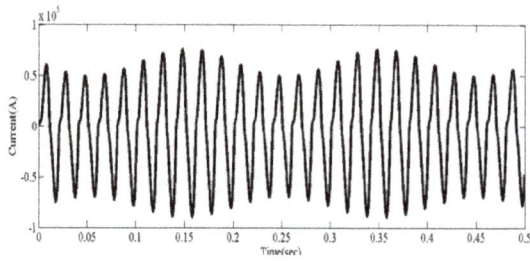

Figure 7. Arc current

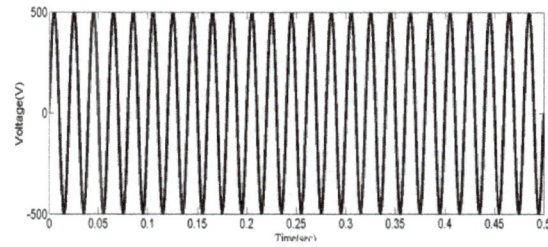

Figure 8. Source voltage after compensation

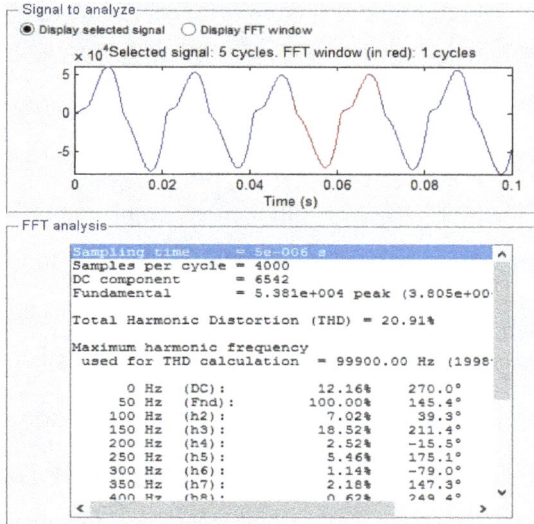

Figure 9. THD of current before compensation

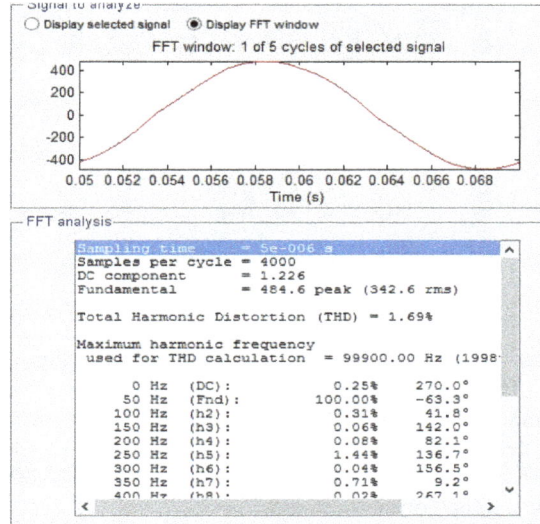

Figure 10. THD of current after compensation

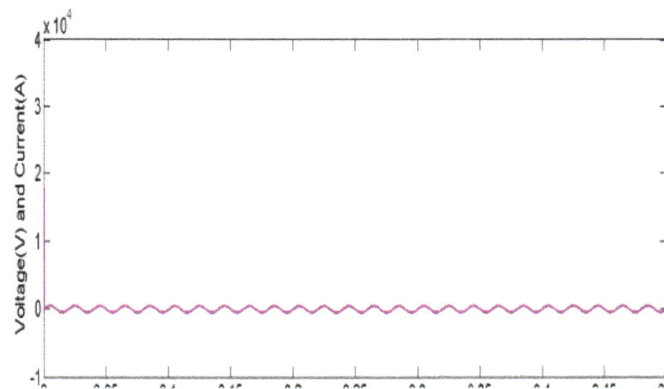

Figure 11. Source voltage and current

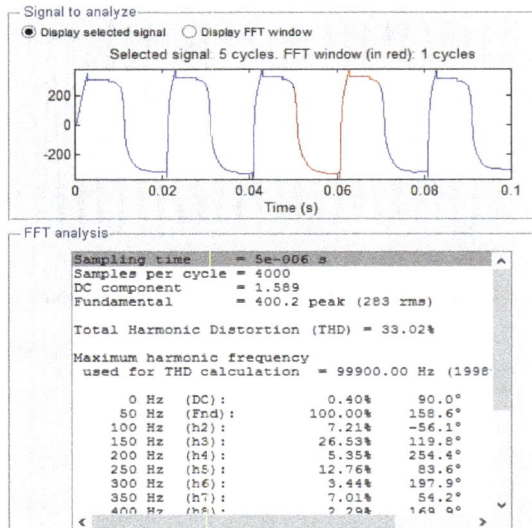

Figure 12. THD of voltage before compensation

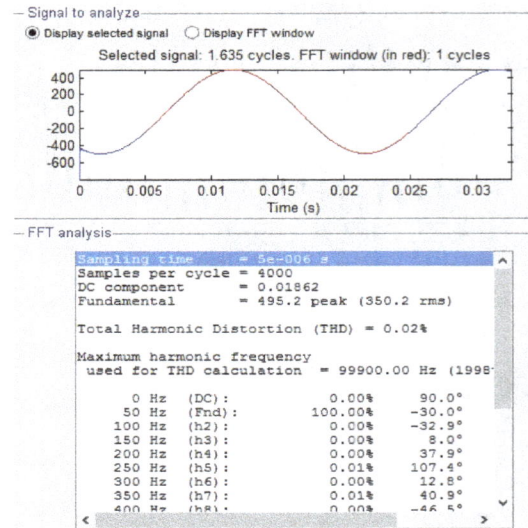

Figure 13. THD of voltage after compensation

5.   CONCLUSION

This paper presents the electric arc furnace model in time domain analysis which was modelled using matlab/simulink and the major power quality issues caused by electric furnace has been studied.The D-STATCOM was used as voltage source inverter which mitigates the power quality issue of voltage flicker created by electric arc furnace. The D-STATCOM control was achieved using Icosϕ control algorithm which has been implememented with few modifications to make the controller more effective. From the simulation results, voltage flicker and THD on current were reduced and the power factor correction for arc furnace load was obtained closer to unity.

REFERENCES

[1]   Ambarnath Banerji, Sujit K Biswas, Bhim Singh. DSTATCOM Application of voltage sag caused by dynamic loads in autonomous systems. *International Journal of Power Electronics and Drive Sytem (IJPEDS)*. 2012; 2(2): 232-240.

[2]   Ambarnath Banerji, Sujit K Biswas, Bhim Singh. D-STATCOM control algorithms: a review. *International Journal of Power Electronics and Drive System (IJPEDS)*. 2012; 2(3): 285-296.

[3]   HL Jou. Performance comparisons of the three-phase active power filter algorithms. *IEE Proc-gener. Transm. distrib.*, 1995; 142(6): 646-652.

[4]   G Bhuvaneshwari, Manjula G Nair. Design, simulation and analog circuit implementation of a three phase shunt active filter using the Icosϕ algorithm. *IEEE transactions on power delivery*. 2008; 23(2): 1222-1235.

[5]   Ilango K, Bhargav A, Trivikram A, Kavya PS, Mounika G, Manjula G Nair. Power quality improvement using STATCOM with renewable energy sources. *IICPE*. IEEE. 2012: 1-6.

[6]   S Pushpavalli, A CordeliaSumathy. Hramonic reduction of arc furnaces using D-STATCOM. *IOSRJEN*. 2013; 3(4): 7-14.

[7]   K Anuradha, BP Muni, AD Raj Kumar. Modelling of electric arc furnace and control algorithms for voltage flicker mitigation using D-STATCOM. *IPEMC IEEE*. 2009: 1123 – 1129.

[8]   Tongxin Zheng, Elham B Makram. An adaptive model arc furnace model. *IEEE transactions on power delivery*. 2000; 15(3): 931-939.

12

Adaptive Fuzzy Integral Sliding-Mode Regulator for Induction Motor Using Nonlinear Sliding Surface

Yong-Kun Lu

School of Electronic Information and Automation, Tianjin University of Science and Technology, Tianjin, China

Keyword:

Adaptive fuzzy control term
Induction motor
Integral sliding-mode controller
Nonlinear sliding surface
Speed regulator

ABSTRACT

An adaptive fuzzy integral sliding-mode controller using nonlinear sliding surface is designed for the speed regulator of a field-oriented induction motor drive in this paper. Combining the conventional integral sliding surface with fractional-order integral, a nonlinear sliding surface is proposed for the integral sliding-mode speed control, which can overcome the windup problem and the convergence speed problem. An adaptive fuzzy control term is utilized to approximate the uncertainty. The stability of the controller is analyzed by Lyapunov stability theory. The effectiveness of the proposed speed regulator is demonstrated by the simulation results in comparison with the conventional integral sliding-mode controller based on boundary layer.

Corresponding Author:

Yong-Kun Lu,
School of Electronic Information and Automation,
Tianjin University of Science and Technology,
1038 Dagunan Road, Hexi District, Tianjin Municipality, PR China.
Email: automationcn@126.com

1. INTRODUCTION

Induction motor (IM) has been widely applied in the industrial field owing to its less-maintenance, lower-cost and excellent-reliability. High variable speed performance of induction motor is achieved through field-oriented control (FOC). In field-oriented control (or vector control), the induction motor can be controlled in a manner similar to the control of separately excited DC motor. The major problem of FOC is the sensitivity to large uncertainties which are due to magnetization saturation, temperature variation, load disturbances, etc [1]. In order to improve the performance of speed regulator under uncertainties in mechanical parameters and load torque, many improved speed regulator of FOC were proposed for induction motor drives [1]-[4]. Due to the good robustness, fast dynamics response and easy implementation, the sliding-mode control has been used in the control of induction motor [5]-[7]. But sliding-mode control is suffering from the chattering phenomenon. One effective solution is replacing the sign function by continuous saturation functions [8]. Boundary layer is a popular saturation function at the cost of the increased steady-state tracking error. The investigations on integral sliding-mode controller for induction motor can be found in [9]-[11]. In [9], an integral sliding-mode control strategy using saturation function was used to stabilize speed tracking of each induction motor while synchronizing its speed with the speed of other motors. A sliding-mode controller was presented for sensorless FOC of induction motor with model reference adaptive system in [10], where an integral sliding-mode control using boundary layer was designed. An integral sliding-mode control using boundary layer was adopted for speed controller of induction motor drives with reference model and a Luenberger observer in [11].

However, the windup problem and the convergence speed problem are not discussed in the above integral sliding-mode control strategies for speed regulator of FOC. As mentioned in [12], the integral action may lead to windup problem, and significant overshoot may occur that requires long time for recovery. To eliminate the windup phenomenon for an integral sliding-mode control, the integral action was turned on only when the norm of tracking errors was lower than a predetermined value in [12]. Moreover, it is well-known that the integral action may slow down the convergence speed of tracking error. The derivative action may speed up the convergence speed of tracking error, but as we know the time derivative of mechanical speed (accelerated mechanical speed) is sensitive to the noise and difficult to obtain at present even using improved differentiators such as nonlinear differentiator [13] and sliding mode differentiator [14]. Hence, accelerated mechanical speed is seldom employed in practical speed regulator of FOC. On the other hand, the investigation of fractional-order control has attracted more and more interests. The fractional-order controller is the extension of integer-order controller [15], which introduces extra degrees of freedom. The fractional-order sliding-mode control were discussed in [16] and [17]. A fractional-order integral sliding-mode flux observer was provided to estimate the d- and q-axis fluxes in the stationary reference frame for sensorless vector controlled induction motors in [16]. A fractional-order sliding-mode control scheme based on parameters auto-tuning for the velocity control of permanent magnet synchronous motor was proposed in [17].

The main contribution of this paper lies in the following three aspects: (1) An adaptive fuzzy sliding-mode controller is proposed and successfully applied to the speed regulator of induction motor. (2) Combining the conventional integral sliding surface with fractional-order integral, a nonlinear sliding surface is proposed for the integral sliding-mode speed regulator, which can overcome the windup phenomenon and speed up convergence. (3) The adaptive fuzzy control term based on the nonlinear sliding surface is applied to approximate the uncertainty.

2.   DYNAMIC MODEL OF INDUCTION MOTOR

The mathematics model of an induction motor can be written in the rotor rotating reference frame (d-q) [10] as follows:

$$\frac{d\omega_r}{dt} = \rho\psi i_q - \beta T_L - \alpha\omega_r$$

$$\frac{d\psi}{dt} = -a\psi + aL_m i_d$$

$$\frac{di_q}{dt} = -\delta i_q - \upsilon\omega_r\psi - \omega i_d + bu_q \tag{1}$$

$$\frac{di_d}{dt} = -\delta i_d + \upsilon a\psi + \omega i_q + bu_d$$

Where $\rho = n_p L_m/(J_m L_r)$, $a = R_r/L_r$, $b = 1/(L_\sigma L_S)$, $\alpha = B_m/J_m$, $\beta = 1/J_m$, $\upsilon = L_m/(L_\sigma L_S L_r)$, $\delta = (R_S L_r^2 + R_r L_m^2)/(L_\sigma L_S L_r^2)$; ω_r and ω are the rotor mechanical speed and the synchronous speed; ψ is the rotor flux; L_m is the mutual inductance; $L_\sigma = 1 - L_m^2/(L_r L_S)$ is the motor leakage inductance; i_d and i_q are the d, q-axis stator currents; R_r and L_r are the rotor resistance and inductance; R_S and L_S are the stator resistance and inductance; n_p is the number of pole pairs; u_d and u_q are the d, q-axis stator voltages; T_L is the external load torque; J_m and B_m are the mechanical inertia of moment and the damping torque coefficient; electromagnetic torque of an induction motor is defined as:

$$T_e = n_p L_m \psi i_q/L_r \tag{2}$$

From (1) and (2), one has:

$$\frac{d\omega_r}{dt} = -\alpha\omega_r - h + \beta T_e \tag{3}$$

Where $h = T_L/J_m$. Consider the uncertainties in (3), one gets:

$$\frac{d\omega_r}{dt} = -(\alpha + \Delta\alpha)\omega_r - (h + \Delta h) + (\beta + \Delta\beta)T_e \tag{4}$$

Where $\Delta\alpha$, Δh and $\Delta\beta$ are the time-varying value of α, h and β, respectively.

The speed tracking error is defined as:

$$e(t) = \omega_r^*(t) - \omega_r(t) \tag{5}$$

Where ω_r^* is the speed reference.

The time derivative of Equation (5) is:

$$\frac{de(t)}{dt} = -ae(t) - \beta u(t) + \xi(t) + \eta(t) \tag{6}$$

Where $\xi(t) = a\omega_r^*(t) + \dfrac{d\omega_r^*(t)}{dt} + h(t)$, uncertainty term $\eta(t) = \Delta a\omega_r(t) - \Delta\beta T_e(t) + \Delta h(t)$, $u = T_e$.

The control objective is to find a speed regulator using adaptive fuzzy sliding-mode controller in rotor flux oriented reference frame for the tracking of speed in presence of model uncertainty.

The overall block diagram for a direct field-oriented induction motor drive is shown in Figure 1, which consists of an induction motor (IM), a SPWM voltage source inverter, two current controllers, two coordinate translators, a current model, and a speed regulator using adaptive fuzzy sliding-mode controller based on a novel nonlinear sliding surface.

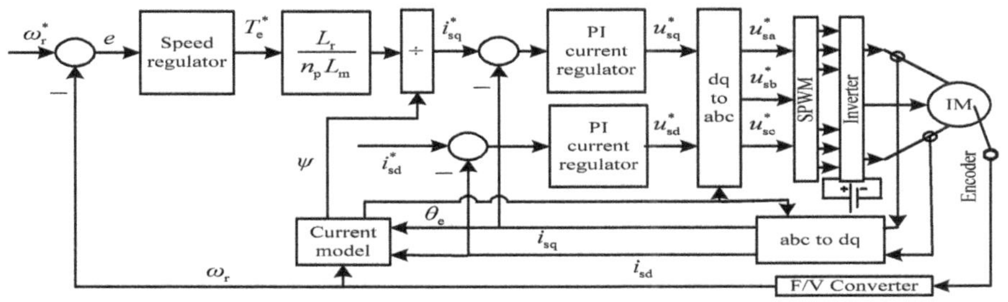

Figure 1. Overall block diagram for a direct field-oriented induction motor drive

3.   DESIGN OF ADAPTIVE FUZZY SLIDING-MODE CONTROLLER

The nonlinear sliding surface can be defined as:

$$S = \begin{cases} e + a_1 e_1 & |e| \le \gamma_1 \\ e + a_2 e_2 & \gamma_1 < |e| \le \gamma_2 \\ e & |e| > \gamma_2 \end{cases} \tag{7}$$

Where a_1, a_2, γ_1 and γ_2 are positive design parameters, $e_1 = \int_0^t e(\tau)d\tau$, $e_2 = {}_0 D_t^\nu(e)$, ${}_0 D_t^\nu(e)$ is the fractional-order integral operator, $-1 < \nu < 0$, the conventional integral sliding surface is used in the small speed tracking error interval to eliminate the windup phenomenon.

According to the exponential reaching law:

$$\frac{dS}{dt} = -fS - K\mathrm{sat}(S/\varphi) \tag{8}$$

Where f and K are positive design parameters, $K \ge |\eta(t)|$. The control law can be designed as:

$$u = \begin{cases} J_m[\xi + u_f + fS + K\text{sat}(S/\varphi)] + (J_m - B_m)e & |e| \le \gamma_1 \\ J_m[\xi + u_f + fS + K\text{sat}(S/\varphi)] + J_{m\,0}D_t^{\nu+1}e - B_m e & \gamma_1 < |e| \le \gamma_2 \\ -B_m e + J_m\xi + J_m u_f + fJ_m S + KJ_m\text{sat}(S/\varphi) & |e| > \gamma_2 \end{cases} \quad (9)$$

Where u_f is the adaptive fuzzy control term to approximate the uncertainty term, $\text{sat}(\cdot)$ is the saturation function defined as:

$$\text{sat}(S/\varphi) = \begin{cases} S/\varphi & |S| \le \varphi \\ \text{sgn}(S) & |S| > \varphi \end{cases} \quad (10)$$

Where $\text{sgn}(\cdot)$ is the sign function, φ is the width of boundary layer which can reduce the chattering phenomenon.

The fractional-order derivative control term $J_{m\,0}D_t^{\nu+1}e$ in (9) is used to speed up convergence of speed tracking error. The approximation of fractional-order derivate and integral plays an important role in the fractional-order control. We adopt the integer-order model to approximate the fractional-order derivate and integral in a suitable frequency interval [18]. The fractional-order derivative used in the proposed controller is not sensitive to the noise beyond the selected frequency interval.

The fuzzy input variables of the adaptive fuzzy control term [19], [20] are S and e. By using the singleton fuzzification, product inference engine and center average defuzzification, the adaptive fuzzy control term is given as:

$$u_f = \boldsymbol{b}^T \boldsymbol{w} = \sum_{j=1}^{m} \frac{\prod_{i=1}^{n} \mu_{F_{ij}} \cdot \hat{u}_j}{\sum_{j=1}^{m} \prod_{i=1}^{n} \mu_{F_{ij}}} \quad (11)$$

Where $\boldsymbol{b} = [\hat{u}_1, \hat{u}_2, \ldots, \hat{u}_m]^T$ is the consequent parameter vector, $\boldsymbol{w}$ is the vector of fuzzy basis functions, $n = 2$, $\mu_{F_{ij}}$ are the membership functions of input variables, $\hat{u}_j$ is the point in output space of the fuzzy system at which the membership function of output variable achieves its maximum value, m is the number of fuzzy rules.

The parameter vector is adapted according to the following updating law:

$$\frac{d\boldsymbol{b}}{dt} = \begin{cases} rS\boldsymbol{w} & ((\|\boldsymbol{b}\| < M_1) \quad \text{or} \quad (\|\boldsymbol{b}\| = M_1 \quad \text{and} \quad S\boldsymbol{b}^T\boldsymbol{w} \le 0)) \\ 0 & \text{others} \end{cases} \quad (12)$$

Where r and M_1 are the positive design parameters.

4. STABILITY ANALYSIS

The optimal parameter vector is defined as:

$$\boldsymbol{b}_0 = \arg\min_{\boldsymbol{b}\in\Omega}[\sup_{\|\boldsymbol{x}\|\le N_1} |\eta(\boldsymbol{x}) - u_f|] \quad (13)$$

And λ is defined as the minimal approximation error.

Choose the Lyapunov functions as:

$$V_1 = \frac{1}{2}S^2(t) \quad (14)$$

$$V_2 = \frac{1}{2}S^2(t) + \frac{1}{2r}\boldsymbol{q}^T(t)\boldsymbol{q}(t) \quad (15)$$

Where $\boldsymbol{q} = \boldsymbol{b} - \boldsymbol{b}_0$.

The derivative of Equation (14) with respect to time is:

$$\frac{dV_1}{dt} = S\frac{dS}{dt} = S(-fS - K\mathrm{sat}(S/\varphi) + \eta(t)) \tag{16}$$

If $|S| > \varphi$, then:

$$\frac{dV_1}{dt} = S(-fS - K\mathrm{sat}(S/\varphi) + \eta(t)) \le -fS^2 - K|S| + |\eta(t)||S| \tag{17}$$

Thus, if the condition of $K \ge |\eta(t)|$ is satisfied, $\dfrac{dV_1}{dt} \le 0$ holds, and $\dfrac{dV_1}{dt} = 0$ only when $S = 0$.

On the other hand, If $|S| \le \varphi$, considering (13), the derivative of Equation (15) with respect to time is:

$$\frac{dV_2}{dt} = S\frac{dS}{dt} + \frac{1}{r}\boldsymbol{q}^{\mathrm{T}}\frac{d\boldsymbol{b}}{dt} \tag{18}$$

From (10), (11), (12) and (18), then:

$$\frac{dV_2}{dt} = S\frac{dS}{dt} + \frac{1}{r}\boldsymbol{q}^{\mathrm{T}}\frac{d\boldsymbol{b}}{dt} = S[-fS - K\mathrm{sat}(S/\varphi) + \lambda] = S[-fS - KS/\varphi + \lambda] \tag{19}$$

If the adaptive fuzzy control term is properly designed, λ is sufficiently small, then $\dfrac{dV_2}{dt} \le 0$ holds,

and $\dfrac{dV_2}{dt} = 0$ only when $S = 0$. That means Lyapunov function V_2 will decrease gradually and the sliding surface will converge to zero. If the system of sliding surface is stable, the speed tracking error will converge to zero.

5.   SIMULATION RESULTS

Simulations are carried out using the Simulink package of MATLAB. The overall control structure for the simulation is shown in Figure 1. The specifications and nominal parameters of motor operated using direct rotor field orientation are given in Table 1 [1].

Table 1. Specifications and Nominal Parameters of an Induction Motor

Motor parameter	Value
Output power (HP)	50
Rated voltage (V)	460
Number of pole pairs (P)	2
Rated frequency (Hz)	60
Stator resistance(Ω)	0.087
Rotor resistance (Ω)	0.228
Stator inductance (mH)	35.5
Rotor inductance (mH)	35.5
Mutual inductance (mH)	34.7
Mechanical inertia of moment(kg•m^2)	2
Damping torque coefficient(N•m•s)	0.2

The operating sequences are described as follows. The initial load torque is constant (0N•m). After the initial constant speed reference of 90rad/s from time t=0 to 0.1s. From time t=0.1 to 0.25s, the speed reference is increased linearly from 90 to 120rad/s, and then from t=0.6 to 0.9s speed reference is decreased from 120 to 90rad/s. At time t=1.1s constant load torque (190N•m) is applied.

The values of mechanical inertia of moment J_m and damping torque coefficient B_m are 0.831kg•m^2 and 0.5N•m•s during the simulation, i.e., there are uncertainties in the mechanical parameters. Simulation tests have been performed in order to compare the dynamic performance of the proposed speed regulator with the conventional integral sliding-mode controller based on boundary layer, i.e., the proposed controller without the nonlinear sliding surface of Equation (7) and the adaptive fuzzy control term of Equation (11).

In the frequency domain, the fractional-order derivative of $_0D_t^{\nu+1}e$ can be expressed as $s^{\nu+1}$, where s is the Laplace variable. Figure 2 and Figure 3 show the bode diagram of the fractional-order derivative $s^{0.2}$ in the simulation and the bode diagram of the integer-order derivative s.

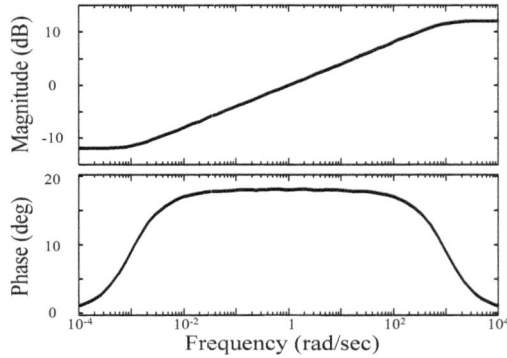

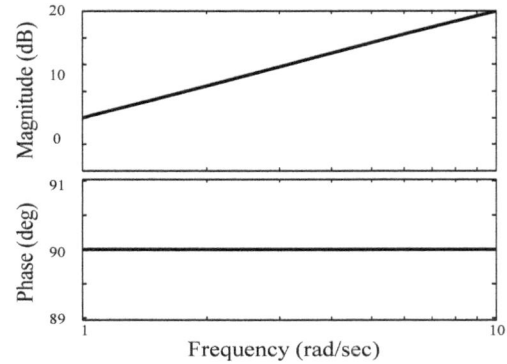

Figure 2. Bode diagram of $s^{0.2}$ in the simulation   Figure 3. Bode diagram of s

The design parameters of the proposed speed regulator are $\nu = -0.8$, $\varphi = 1.5$, $\gamma_1 = 1$, $\gamma_2 = 5$, $r = 100$, $M_1 = 20$, $a_1 = 1$, $a_2 = 1$, $f = 1$, $K = 100$ and the fuzzy membership functions of e are designed as:

$\mu_{F_{11}} = \min(1, \max(0, 1 - (4e + 6)/3))$; $\mu_{F_{12}} = \max(0, \min(1 + (4e + 3)/3, 1 - (8e + 6)/3))$;

$\mu_{F_{13}} = \max(0, \min(1 + (8e + 3)/3, 1 - (8e + 3)/3))$; $\mu_{F_{14}} = \max(0, \min(1 + 8e/3, 1 - 8e/3))$;

$\mu_{F_{15}} = \max(0, \min(1 + (8e - 3)/3, 1 - (8e - 3)/3))$; $\mu_{F_{16}} = \max(0, \min(1 + (8e - 6)/3, 1 - (4e - 3)/3))$;

$\mu_{F_{17}} = \min(1, \max(0, 1 + (4e - 6)/3))$.

The fuzzy membership functions of S are the same as those of e. The sliding surface of the compared controller is selected as $S = e + e_1$.

Figure 4 shows the desired motor speed (Dash-dot line), the rotor speed based on the compared controller (Dashed line) and the rotor speed based on the proposed controller (Solid line). It is clear that the rotor speed performance of the proposed controller is better than that of the compared controller after the step change of external load torque.

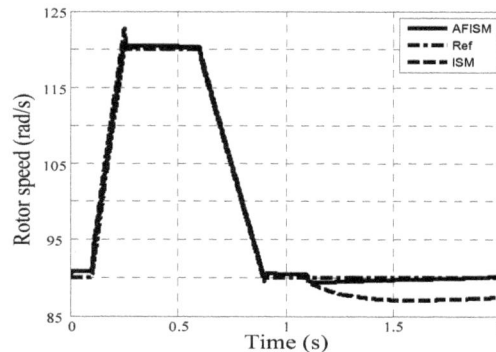

Figure 4. Reference speed and rotor speed response

The performances of the motor torque are illustrated by Figure 5 and Figure 6. It is seen that the motor torques are within reasonable ranges in Figure 5-6.

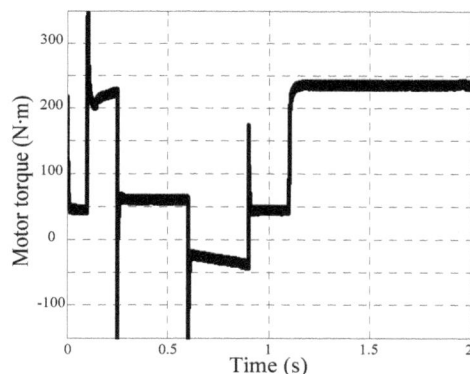

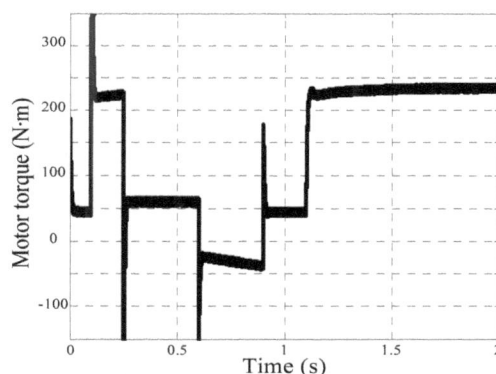

Figure 5. Torque response of the proposed controller   Figure 6. Torque response of the compared controller

The simulation results reveal that the presented method has better tracking performance than the conventional integral sliding-mode controller based on boundary layer under uncertainties in the mechanical parameters and load torque.

6. CONCLUSION

In this paper, an adaptive fuzzy sliding-mode vector control has been presented for speed regulator of induction motor. It is proposed as a sliding-mode controller which has a nonlinear sliding surface to overcome the windup phenomenon of conventional integral sliding-mode speed controller strategy and speed up convergence by fractional-order derivative control term which is not sensitive to the noise beyond the selected frequency interval. Moreover, the proposed sliding-mode controller incorporates a fractional-order adaptive fuzzy control term based on the nonlinear sliding surface to approximate the uncertainty. Then the closed loop stability of the presented design has been proved by Lyapunov stability theory. Finally, by means of simulation examples, it has been shown that the proposed control method improves tracking performance of speed in comparison with the conventional integral sliding-mode controller based on boundary layer in presence of external load disturbance and mechanical parameter variations.

ACKNOWLEDGEMENTS

This work is supported by the Science and Technology Development Foundation of the Higher Education Institutions of Tianjin Municipality of China (No. 20130722).

REFERENCES

[1] O Barambones, AJ Garrido. An Adaptive Variable Structure Control Law for Sensorless Induction Motors. *European Journal of Control*. 2007; 13: 382-392.

[2] ZMS Barbary. DSP Based Vector Control of Five-Phase Induction Motor Using Fuzzy Logic Control. *International Journal of Power Electronics and Drive System*. 2012; 2: 192-202.

[3] ESD Santana, E Bim, WCD Amaral. A Predictive Algorithm for Controlling Speed and Rotor Flux of Induction Motor. *IEEE Transactions on Industrial Electronics*. 2008; 55: 4398-4407.

[4] A Mishra, P Choudhary. Artificial Neural Network Based Controller for Speed Control of an Induction Motor Using Indirect Vector Control Method. *International Journal of Power Electronics and Drive System*. 2012; 2: 402-408.

[5] ZH Akpolat, H Guldemir. Trajectory Following Sliding Mode Control of Induction Motors. *Electrical Engineering*. 2003; 85: 205-209.

[6] M Hajian, GR Markadeh, J Soltani, *et al.* Energy Optimized Sliding-Mode Control of Sensorless Induction Motor Drives. *Energy Conversion and Management*. 2009; 50: 2296-2306.

[7] JB Oliveira, AD Araujo, SM Dias. Controlling the Speed of a Three-Phase Induction Motor Using a Simplified Indirect Adaptive Sliding Mode Scheme. *Control Engineering Practice*. 2010; 18: 577-584.

[8] M Sulaiman, FA Patakor, Z Ibrahim. New Methodology for Chattering Suppression of Sliding Mode Control for Three-Phase Induction Motor Drives. *WSEAS Transactions on System and Control*. 2014; 9: 1-9.

[9] DZ Zhao, CW Li, J Ren. Speed Synchronization of Multiple Induction Motors with Total Sliding Mode Control. *Systems Engineering-Theory and Practice*. 2009; 29: 110-117.

[10] G Abdelmadjid, BM Seghir, S Ahmed, *et al.* Sensorless Sliding Mode Vector Control of Induction Motor Drives. *International Journal of Power Electronics and Drive System.* 2012; 2: 277-284.

[11] SE Rezgui, H Benalla. New Robust and Mechanical Sensorless Scheme for SVM Inverter Fed Induction Motor Drive Using Variabe Structure Controllers and MRAS. *Arabian Journal for Science and Engineering.* 2013; 38: 1449-1458.

[12] H Joe, M Kim, SC Yu. Second-Order Sliding-Mode Controller for Autonomous Underwater Vehicle in the Presence of Unknown Disturbances. *Nonlinear Dynamics.* 2014; 78: 183-196.

[13] XH Wang, B Shirinzadeh. High-Order Nonlinear Differentiator and Application to Aircraft Control. *Mechanical Systems and Signal Processing.* 2014; 46; 227-252.

[14] H Alwi, C Edwards. An Adaptive Sliding Mode Differentiator for Actuator Oscillatory Failure Case Reconstruction. *Automatica.* 2013; 49: 642-651.

[15] DJ Wang, XL Gao. H∞ Design with Fractional-Order Controllers. *Automatica.* 2012; 48: 974-977.

[16] YH Chang, CI Wu, HC Chen, *et al.* Fractional-Order Integral Sliding-Mode Flux Observer for Sensorless Vector-Controlled Induction Motors. *American Control Conference. ACC 2011.* 2011: 190-195.

[17] B Zhang, Y Pi, Y Luo. Fractional Order Sliding-Mode Control Based on Parameters Auto-Tuning for Velocity Control of Permanent Magnet Synchronous Motor. *ISA Transactions.* 2012; 51: 649-656.

[18] W Krajewski, U Viaro. A Method for the Integer-Order Approximation of Fractional-Order Systems. *Journal of the Franklin Institute.* 2014; 351: 555-564.

[19] YJ Liu, YX Li. Adaptive Fuzzy Output-Feedback Control of Uncertain SISO Nonlinear Systems. *Nonlinear Dynamics.* 2010; 61: 749-761.

[20] L Li, XZ Zhang. Indirect Adaptive Fuzzy Sliding-Mode Control for Induction Motor Drive. *Journal of Electrical Systems.* 2011; 7: 412-422.

13

Model of Pulsed Electrical Discharge Machining (EDM) using RL Circuit

Ade Erawan Minhat*, Nor Hisham Hj Khamis**, Azli Yahya*,
Trias Andromeda*, Kartiko Nugroho***
* Department of Electronic and Computer Engineering, Universiti Teknologi Malaysia
** Department of Communication Engineering, Universiti Teknologi Malaysia
*** Department of Biotechnology and Medical Engineering, Universiti Teknologi Malaysia

ABSTRACT

Keyword:

Electrical Discharge Machining
Pulse Width Modulation
Gap current
Gap voltage

This article presents a model of pulsed Electrical Discharge Machining (EDM) using RL circuit. There are several mathematical models have been successfully developed based on the initial, ignition and discharge phase of current and voltage gap. According to these models, the circuit schematic of transistor pulse power generator has been designed using electrical model in Matlab Simulink software to identify the profile of voltage and current during machining process. Then, the simulation results are compared with the experimental results.

Corresponding Author:

Ade Erawan Minhat,
Department of Electronic and Computer Engineering,
Faculty of Electrical Engineering,UniversitiTeknologi Malaysia,
81310, Johor Bahru, Johor.
Email: adeerawan@gmail.com

1. INTRODUCTION

Electrical Discharge Machining (EDM) is a machining process that enables noncontact drill via electrochemical effects irrespective of the hardness of the workpiece (see Figure 1). In EDM process, pulse power generator is required in order to obtain the discharge spark. The efficiency of production is depending on the performance of the pulse power generator. Control servo is used to control the space gap between electrode and workpiece. In creating the spark discharge, a current flow from the electrode through a dielectric fluid due to the gap distance between electrode and workpiece is reduced to a very small clearance approximately 10 to 50 microns [1, 2]. Electrical energy from the spark is converted into heat energy, then builds up the workpiece temperature and melts the area on its surface. The working pulse power generator is an important role in affecting the material removal rate (MRR) and the properties of the machined surface [3, 4]. The filtration system is used to maintain the dielectric fluid and flush out the eroded gap particles. This article presents the pulse phase in the EDM process due to improve in machining parameter. In order to prove the theoretical more clearly is determine by performing the simulation and experimental studies.

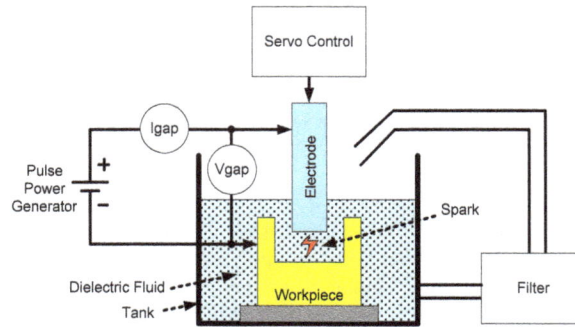

Figure 1. EDM System

2. EDM POWER GENERATOR

Generally, EDM power generator is configured by two important parts known as power supply and pulse generator is shown in Figure 2 [5]. There are several of power supply can be used, such as linear power supply and switching mode power supply (SMPS). Base on the power consumption cost issue, higher material removal rate and good surface finish in EDM parameter, the study is focused on the switching power supply [6]. By using the SMPS topology, the configuration has a high efficiency and high performance [1], [7]-[8].

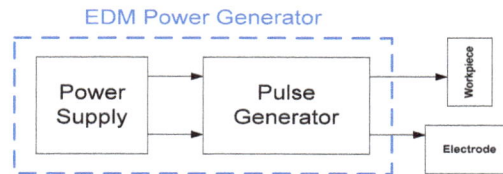

Figure 2. Block diagram for EDM Power Generator

Pulse generator is divided into two types. There are relaxation (resistance-capacitance) generator and transistor pulse generator. The relaxation circuit type of EDM pulse power generator create pulses through the capacitor charge and discharge behavior. Discharge energy is determined by the used capacitance and by the stray capacitance that exists between electrode and workpiece. The electrical sparks are created from the released charges of capacitor.

The transistor pulse generator is widely used in conventional EDM and provides a higher MRR due to its high discharge energy [9]-[11]. Moreover, the pulse duration and discharge current can be arbitrarily changed depending on the required machining characteristics. The transistor pulse generator generates a rectangular pulse discharges by controlling the current or voltage source. By changing the duty cycle, pulse width modulation is used to control the transistor states. To ensure a constant processing, the MOSFET transistor is used as a switch to control the output pulse power as shown in Figure 3.

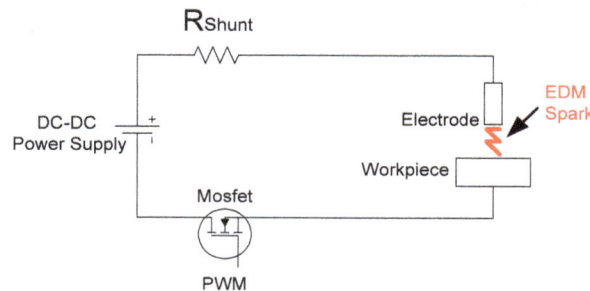

Figure 3. Transistor type of EDM Pulse Power Generator

3.   MODELLING EDM SYSTEM

In this study, a modelof EDM pulse power generator was developed to investigate the pulse profile during EDM process.Based on Figure 3, the schematic circuit of EDM pulse generator has been developed and the mathematical model has been proved by the derived equation.In this schematic design, DC power source as an input source is connected to resistor R_1 (load). Then connected to the gap model between electrode and workpiece which is consisting of R_{ig}, R_{dis} and L_{dis}. To get pulse signal at the output side, it is connected to the MOSFET.Basically there are three phases in the pulse EDM is known as the initial phase, the ignition phase and discharge phase.

3.1. Initial Phase

As can been seen from Figure 4, the schematic circuitof EDM pulse generator and the gap model has been designed. In the initial phase of EDM process, the gap is in open circuit state while switch S_1 is off.In this condition, the output voltage is equal to V_{gap} and current gap is zero.This is occur when the position of the electrode and the workpiece is far or non-discharge.

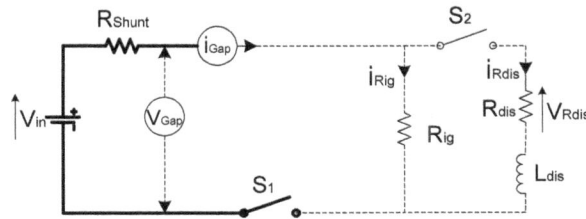

Figure 4. The circuit in ignition phase condition

By applying Kirchhoff's voltage law. The voltage gap is in open circuit voltage state can be expressed as follows.

$$V_{in} = V_{R_{shunt}} + V_{gap} \tag{1}$$

$$V_{gap} = V_{in} - V_{R_{shunt}} = V_{oc} \tag{2}$$

When the circuit is not formed in a closed-loop network, then no current through in the circuit.

$$i_{gap} = \frac{V_{gap}}{R_{shunt} + R_{ig}} = 0 \tag{3}$$

3.2. Ignition Phase

In the ignition phase, a strong electric field is established between electrode and workpiece.Due to the attractive force of the electric field, there is created an ionization path through the dielectric. During the process, if ignition delay time is too long, this means the circuit is in open circuit and if the ignition delay time is too short, this means the circuit is a short circuit. Both cases are abnormal. It is important keep the ignition delay time to be a constant. From Figure 5, the switch S_1 is turn on and S_2 is turn off. The circuit is formed in a closed loop network. The gap voltage is refers to the voltage through resistor R_{ig} which is become a voltage divider between resistors R_{ig} and R_{shunt}.

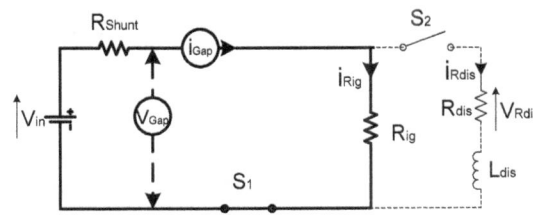

Figure 5. The circuit in ignition phase condition

Applying Kirchhoff's current law,

$$i_{gap} = i_{R_{ig}} + i_{R_{dis}} \tag{4}$$

When $i_{R_{dis}}$ is zero, current gap during ignition phase can be expressed as follows,

$$i_{gap} = i_{R_{ig}} \tag{5}$$

According to Figure 5, thecircuit is formed in a closed-loop network. The gap voltage is the difference between V_{in} and voltage across R_{ig}. By applying Kirchhoff's voltage law, gap voltage can be expressed as follows,

$$V_{in} = V_{R_{shunt}} + V_{gap} \tag{6}$$

$$V_{in} = i_{gap}R_{shunt} + V_{gap} \tag{7}$$

$$V_{gap} = V_{in} - i_{gap}R_{shunt} \tag{8}$$

From Equation (8), the gap voltage can be express as the voltage divider ruleduring the ignition phase,

$$V_{gap} = \frac{R_{ig}}{R_{shunt}+R_{ig}} V_{in} \tag{9}$$

3.3. Discharge Phase

During the discharge phase, it is initiated by moving the electrode very closeto the workpiece. A plasma channel has been form due to ionization of dielectric. Due to the spark gap, voltage drops and current rises abruptly which forms the crater at spot of discharge on the workpiece.

As evident in Figure 6, both of switch S_1 and switch S_2 is turn ON. Switch S_1 has been used due to control the main pulse in pulse generator such duty cycle, time ON and time OFF. Whereas, switch S_2 used to control the transient current and voltage drop during the discharge phase. In order to get current gap i_{gap}, it is obtained by combination between current through resistor R_{ig} and current at $i_{R_{dis}}$.

Refer to the gap model in Figure 6, it consist an inductance L_{dis} connected in series with a resistance R_{dis} and parallel with resistance R_{ig}. The transient time of current and voltage during the discharge phase is determined by the relationship between the inductance L_{dis} and the resistance R_{dis}. The fixed value resistanceR_{dis} and larger the inductance L_{dis}, the slower will be the transient time. However, for a fixed value inductance L_{dis}, by increasing the resistance value R_{dis}, fast transient time and therefore the time constant of the circuit becomes shorter. In general, the voltage will drop to about 20V-30V during discharge time [12].

Then, the process will be repeated to the ignition phase which is both switch S_1 and switch S_2 is turn off. All phases will be repeated until the end of theEDM process.

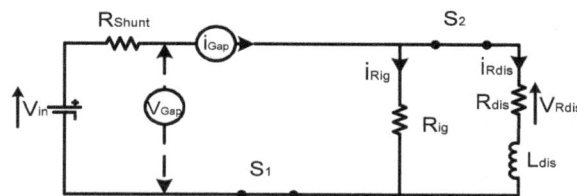

Figure 6. The circuit in discharge phase condition which is switch (S_1) and switch (S_2) is turn ON

In mathematical model, the gap voltage can be expressed as follows.

$$V_{gap} = i_{R_{dis}}R_{dis} + L_{dis}\frac{di_{R_{dis}}}{dt} \tag{10}$$

$$V_{gap} - i_{R_{dis}}R_{dis} - L_{dis}\frac{di_{R_{dis}}}{dt} = 0 \tag{11}$$

$$V_{gap} - i_{R_{dis}}R_{dis} = L_{dis}\frac{di_{R_{dis}}}{dt} \tag{12}$$

After adjusted,

$$\frac{dt}{L_{dis}} = \frac{di_{R_{dis}}}{V_{gap} - i_{R_{dis}}R_{dis}} \tag{13}$$

Integrating both the equations,

$$\int_0^t \frac{dt}{L_{dis}} = \int_0^i \frac{di_{R_{dis}}}{V_{gap} - i_{R_{dis}}R_{dis}} \tag{14}$$

$$\frac{t}{L_{dis}} = \int_0^i \frac{di_{R_{dis}}}{V_{gap} - i_{R_{dis}}R_{dis}} \tag{15}$$

By using assumption,

$$z = V_{gap} - i_{R_{dis}}R_{dis} \tag{16}$$

$$\frac{dz}{di_{dis}} = -R_{dis} \tag{17}$$

$$di_{R_{dis}} = -\frac{dz}{R_{dis}} \tag{18}$$

So, $-\frac{R_{dis}t}{L_{dis}}$ can be expressed as follows.

$$\frac{t}{L_{dis}} = -\frac{1}{R_{dis}}\int_0^i \frac{dz}{z} \tag{19}$$

$$-\frac{R_{dis}t}{L_{dis}} = \int_0^i \frac{dz}{z} \tag{20}$$

By using integration rule,

$$\ln(z) = \int_0^i \frac{dz}{z} \tag{21}$$

The Equation (20), can be expressed as follows,

$$-\frac{R_{dis}t}{L_{dis}} = \ln(z)_0^i \tag{22}$$

$$-\frac{R_{dis}t}{L_{dis}} = \ln(V_{gap} - i_{R_{dis}}R_{dis})_0^i \tag{23}$$

Applying limits, $-\frac{R_{dis}t}{L_{dis}}$ can be expressed as follows,

$$-\frac{R_{dis}t}{L_{dis}} = \ln(V_{gap} - i_{R_{dis}}R_{dis}) - \ln(V_{gap}) \tag{24}$$

$$-\frac{R_{dis}t}{L_{dis}} = \ln\left(\frac{V_{gap} - i_{R_{dis}}R_{dis}}{V_{gap}}\right) \tag{25}$$

Taking antilog on both sides in Equation (25),

$$e^{-\frac{R_{dis}t}{L_{dis}}} = \frac{V_{gap} - i_{R_{dis}}R_{dis}}{V_{gap}} \tag{26}$$

$$V_{gap}e^{-\frac{R_{dis}t}{L_{dis}}} = V_{gap} - i_{R_{dis}}R_{dis} \qquad (27)$$

The current $i_{R_{dis}}$ flow through inductance L_{dis} in series to resistance, R_{dis} can be expressed as follows.

$$i_{R_{dis}} = \frac{V_{gap}}{R_{dis}}\left(1 - e^{-\frac{R_{dis}t}{L_{dis}}}\right) \qquad (28)$$

Then, the current gap can be obtained as follows.

$$i_{gap} = i_{R_{ig}} + i_{R_{dis}} \qquad (29)$$

In using Equation (4), the current gap in discharge condition is,

$$i_{gap} = \frac{V_{gap}}{R_{ig}} + \frac{V_{gap}}{R_{dis}}\left(1 - e^{-\frac{R_{dis}t}{L_{dis}}}\right) \qquad (30)$$

$$i_{gap} = V_{gap}\left[\frac{1}{R_{ig}}\left(1 - e^{-\frac{R_{dis}t}{L_{dis}}}\right) + \frac{1}{R_{dis}}\right] \qquad (31)$$

Using the Kirchhoff law again, V_{in} can be determined by,

$$V_{in} = i_{gap}R_{shunt} + V_{gap} \qquad (32)$$

In this phase $V_{gap} = V_{dis}$, the discharge voltage can be represented as below,

$$V_{dis} = V_{in} - i_{gap}R_{shunt} \qquad (33)$$

As illustrated in Figure 7, the three phases of EDM pulses has been shown in details. based on the time duration in one period, the initial phase from 0 until t_1, followed by the ignition phase of the t_1 to t_2 and the next phase of the discharge of the t_2 to t_3.

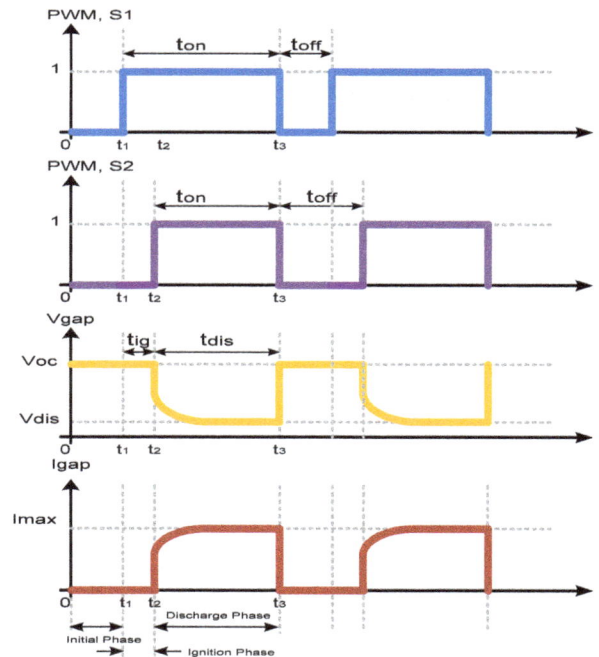

Figure 7. The profile of EDM pulse which is consists switch (S_1), switch (S_2), gap voltage (V_{gap}) and current gap (I_{gap}) versus of time

4.   SIMULATION AND EXPERIMENTAL RESULTS

By using the electrical model in Matlab Simulink software, the simulation process has been conducted. As can be seen in Figure 8, the configuration of the EDM circuit was constructed based on the mathematical model derived.In this simulation, the parameters have been set as the input voltage is 100V, 50 percent duty cycle and 100 microsecond time period. Displayed in Figure 9(a) shows the results obtained from the simulation design is open circuit voltage, V_{oc}=100V, discharge voltage, V_{dis}=28V and current gap, I_{gap}=2.8A.

In the experimental, transistor type of EDM pulse power generator is used to the design. The following input process parameters are used such as input voltage, V_{in}=100V, load resistance, R_{load}=113Ω and copper material for electrode and workpiece. As can be observed in Figure 9(b), the output result shows the open circuit voltage, V_{oc}=95V, discharge voltage, V_{dis}=18V and current gap(current through the load resistance), I_{gap}=0.8A. Comparing the simulation and the experimental results, it is evident that these result are in good agreement with the mathematical model derived.

To analyze the completed result, surface finish of the experimental material were viewed under the OMAX Microscope about100Xmagnification as shown in Figure 10(b) and Figure 10(c). The result shows the diameter hole is about 1 mm with better surface quality. Usually, a small current gap obtained the better surface finish compare with higher current [13].

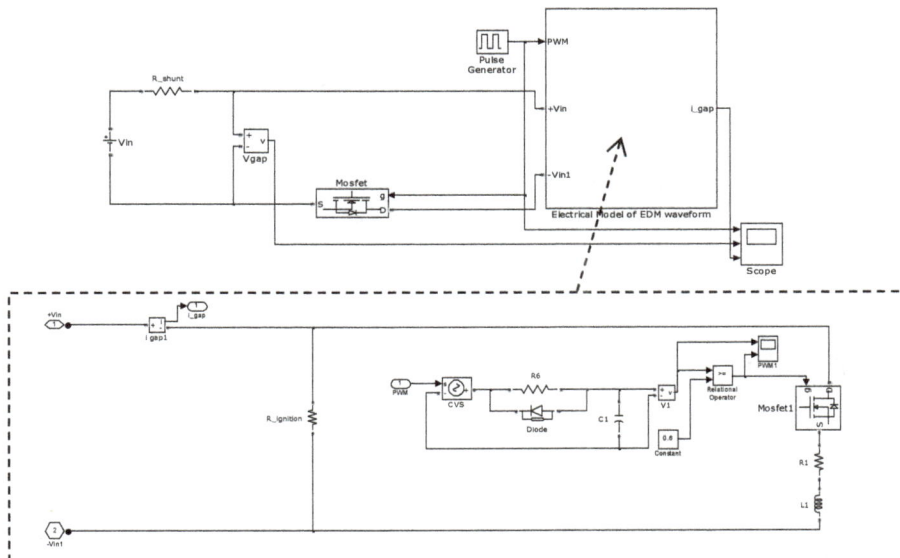

Figure 8. The electrical model of EDM pulse power generator and the configuration of EDM pulses inside the block diagram

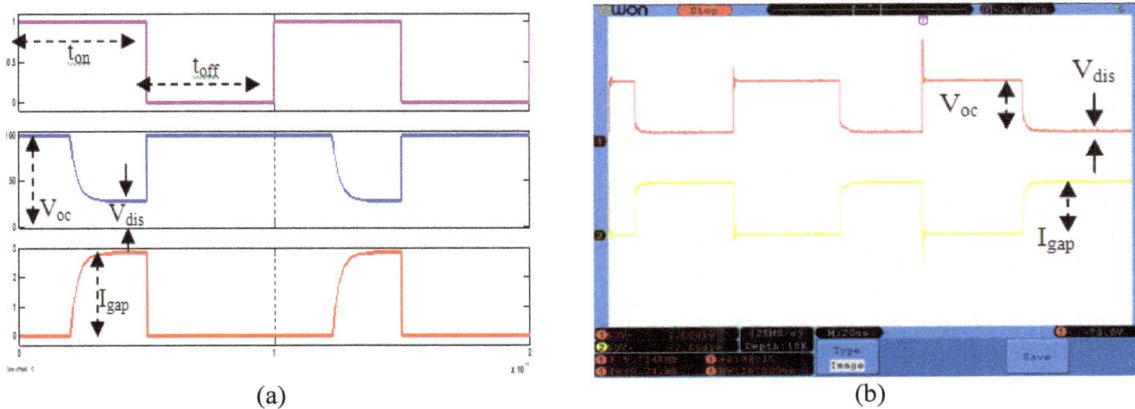

(a)   (b)

Figure 9. (a)The simulation results show the pulse width modulation, voltage and current in the gap. (b)The gap waveform displayed from the experiment (Ch1: Gap Voltage, Ch2: Gap Current)

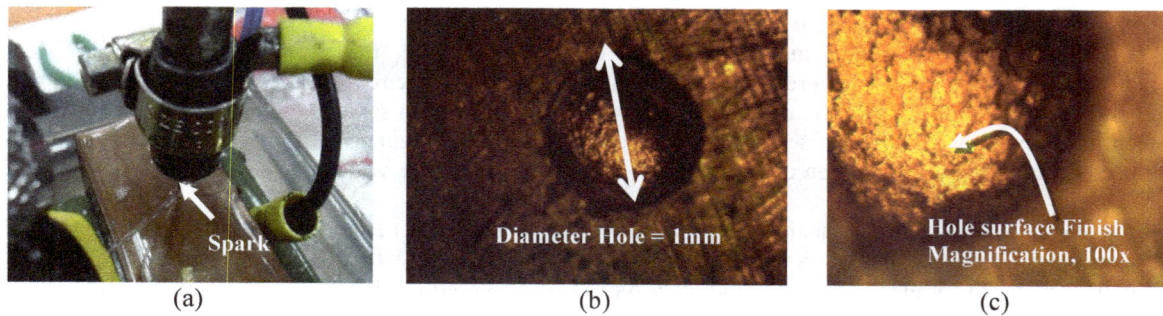

Figure 10. (a) Spark discharge phase; (b) Holes fabricated by the transistor pulse generator; (c) Zooming into the hole surface at 100X magnification

5. CONCLUSION

In conclusion, a new mathematical model of EDM pulses has been presented and implemented successfully. Based on current and voltage gap, there are three mathematical models has been developed such as initial, ignition and discharge phase. Referring to the equations described above, Equation (2) and Equation (3) can be used in an initial phase conditions while Equation (5) and Equation (9) on the ignition phase and Equation (31) and Equation (33) for discharge phase. Mathematical model of EDM pulses as the objective of this study has been achieved. The model has been validated by simulation and experimental result. The performance of the simulation design has been tested and give a good result compared with the theoretical pulse shape.Comparing simulation and experimental result, this mathematical model is applicable to other simulation studies relating to the EDM pulses. This is great theoretical and practical importance for EDM process.

ACKNOWLEDGMENTS

The authors would like to thank the Universiti Teknologi Malaysia (UTM) and Ministry of Education (MoE) Malaysia for financial support through Research University Grant (GUP) with Vote No.05H41.

REFERENCES

[1]   A Yahya, et al. Communication within hardware of Electrical Discharge Machining (EDM) system. Jurnal Teknologi (Special Edition) Sciences & Engineering. 2011; 55(1): 201-212.
[2]   A Yahya. Digital control of an electro discharge machining (EDM) system. Ph.D Thesis, Loughborough University, 2005.
[3]   AE Minhat, et al. Power Generator of Electrical Discharge Machining (EDM) System. Applied Mechanics and Materials. 2014; 554: 638-642.
[4]   Y Tsai, C Lu. Influence of current impulse on machining characteristics in EDM. Journal of Mechanical Science and Technology. 2007; 21: 1617-1621.
[5]   AE Minhat, et al. Model of Pulse Power Generator in Electrical Discharge Machining (EDM) System. Applied Mechanics and Materials. 2014; 554: 613-617.
[6]   A Yahya, et al. Comparison Studies of Electrical Discharge Machining (EDM) Process Model for Low Gap Current. Advanced Materials Research Journal. 2012; 133-440: 650-654.
[7]   Y Yang, et al. Design of pulse power for EDM based on DDS chip AD9851. Mechanic Automation and Control Engineering (MACE), 2010 International Conference on. 2010: 3302-3304.
[8]   AE Minhat, et al. Control Strategy for Electrical Discharge Machining (EDM) Pulse Power Generator. Applied Mechanics and Materials. 2014; 554: 643-647.
[9]   T Andromeda, et al. Predicting material removal rate of Electrical Discharge Machining (EDM) using artificial neural network for high Igap current. Kuantan, 2011.
[10] N Mahmud, et al. Pulse Power Generator Design for Machining Micro-pits on Hip Implant. Jurnal Teknologi (Sciences & Engineering). 2013; 61: 33-38.
[11] M Rahman. A comparative study of transistor and RC pulse generators for micro-EDM of tungsten carbide. International Journal of Precision Engineering and Manufacturing. 2008; 9: 3-10.
[12] W Mysinski. Power supply unit for an electric discharge machine. Power Electronics and Motion Control Conference, 2008. EPE-PEMC 2008. 13th, 2008: 1321-1325.
[13] M Gostimirovic, et al. Effect of electrical pulse parameters on the machining performance in EDM. Indian Journal of Engineering & Materials Sciences. 2012; 18: 411-415.

14

Potential Development of Vehicle Traction Levitation Systems with Magnetic Suspension

A.V. Kireev, N.M. Kozhemyaka, G.N. Kononov
Scientific and Technical Center "PRIVOD-N", Rostov region, Russia

---

Keyword:

Lateral stabilization
Linear motor
Magnetic suspension
Switched reluctance motor
Traction levitation system
Vehicle

ABSTRACT

Below is given the brief analysis of development trend for vehicle traction levitation systems with magnetic suspension. It is presented the assessment of potential development of traction levitation systems in terms of their simplicity. The examples are considered of technical solutions focused on reducing the complexity of transport systems. It is proposed the forecast of their further development.

Corresponding Author:

G.N Kononov,
Scientific and technical center "PRIVOD-N",
Krivoshlykova 4A, Novocherkassk, Rostov region, 346428, Russia
Email: privod-n@privod-n.ru

---

1. INTRODUCTION

Nowadays in Russia due to increase of transport problems the interest in transport technologies based on contactless movement of objects has renewed. The development programs of magnetic-levitation transport providing realization of a number of difficult and expensive projects on design of passenger transport for megalopolises, high-speed transport of distant following, freight transport and systems of conveyor movement based on design of new technology "MagTranCity" have been worked out [1].

Basic components "MagTranCity" are combined use in magnetic levitation poles, lateral stabilization and traction permanent magnets and bulk high-temperature superconductors (flight) of racetrack modules of composite low-temperature superconductors, and the combination of travel levitation tracks as "winding Gram" shorted coils and discrete T-shaped squirrel cells. In world practice the magnetic-levitation transport is not still widely used due to the high cost of construction and insufficiently intensive passenger flow. All range of its transport system's commercial operation based on «Transrapid» technology is limited by the line 30km long connecting Shanghai with international airport Pudong.

At the same time the long-felt need for development of transport systems, having the isolated track with trestle laying of route for transportation of steady passenger traffic, generates a problem for searching of technical solutions allowing us to lower the cost for design and operation of magnetic-levitation transport system.

2. DEVELOPMENT DIRECTIONS OF TRACTION LEVITATION SYSTEMS

From the very beginning of magnetic-levitation transport systems' development it was proposed several various principles of levitation and traction and within each principle there are a lot of variants and modifications. But none of them has become preferable, due to this fact the problem of selection the best variants are still actual [2]. In the objective context the selection criteria is the technical solutions allowing us to reduce the cost for system's development.

From the history of engineering it is known that the only variant having low complexity of new technical system is implemented, survived and selected. The simplicity and related survivability, reliability are reached during development of a new technical object. This regularity is distinctly traced in the development direction of traction levitation systems. Let's give some examples. As is well known the operation of magnetic levitation vehicle generally requires generating the force systems in vertical and horizontal planes. One of them realizes the magnetic suspension of the vehicle in the vertical plane, the other one is a guiding force and the third one ensures the movement in the horizontal plane. In the first development works for creation of each force system rather self-contained units were used. The mechanical configuration of three-functional levitation system, lateral stabilization and traction, presented at figure 1, can be a typical example. At the vehicle bogie frame 1 it is installed the following: suspension electromagnets 2, lateral stabilization magnets 3, stator of linear traction motor 4. П-shaped ferromagnetic rails 6 and 7, and the lines of passive rotor 8 of linear motor are located through the track trestle 5. The attractive forces of electromagnet 2 to ferrorail 6 are used to generate the levitation, in the same way by means of electromagnets 3 and ferrorail 7 the system of guiding forces, preventing the shift of vehicle in the lateral direction, are created. The movement along the track is realized due to interaction forces of linear motor stator 4, installed at the bogie frame of the vehicle, with a passive rotor 8 located at the track. Such mechanical configuration was found to be material-intensive and bulky.

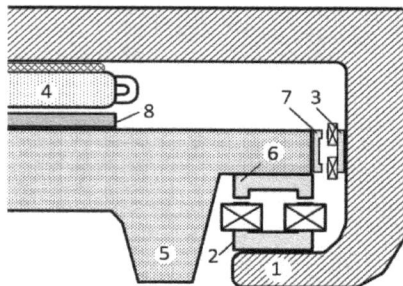

Figure 1. Three-functional mechanical configuration of Traction levitation system

The further development of system was related with attempts to combine the structural elements in the way to merge various functions. The most common merged functions are magnetic suspension and guidance. The example can be the scheme given at Figure 2. At the bogie frame of the vehicle 1 it is installed the Traction levitation module, contained the inductor of linear traction motor 2, electromagnets 3, 4 with L-shaped magnetic conductors used for interaction with track П-shaped ferromagnetic guide rail 5 fixed at the track 6. Between the guide rail branches the inductor of linear traction motor 2 is set. When power supplying at electromagnets 3, 4 the magnetic flux is isolated by means of L-shaped magnetic conductors of electromagnets and magnetic conductors of guide rail 5 that generates lifting and lateral effort of stabilization and in case of supplying the inductor 2 of linear motor, the vehicle starts moving.

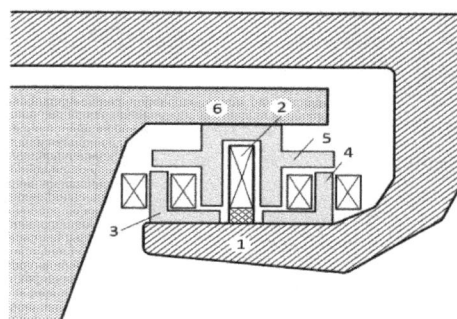

Figure 2. Combined scheme of traction levitation system

The further search of possible increasing of mass-dimensional and power indicators led the designers to the idea of integration the different functions within a single power element. The integration principle was practically realized in the system designed on the basis of three-functional Linear Synchronous Motor (LSM) [3]. LSM design, presented at figure 3, contains the yoke 1; drive winding 2; winding of transverse stabilization 3; traction winding 4; track element 5; anchor tooth 6; track structure 7.

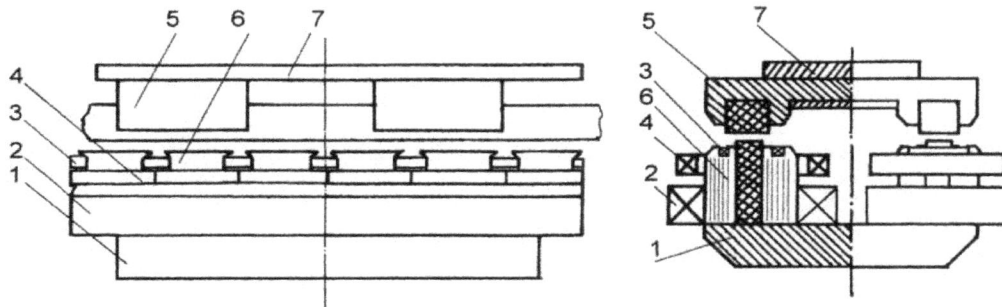

Figure 3. LSM design

The motor inductor consists of U-shaped magnetic conductor in the form of two cores 6 connected by yoke 1. The cores of inductor 6 are divided in longitudinal direction into teeth with three-phase traction winding 4 inside. There is a transverse stabilization winding 3 inside longitudinal slots of the teeth. Coils 2 of drive winding are installed at the inductor's cores; when current supply these coils generate the magnetic flux going through yoke 1, teeth 6 and closing through air gaps at ferromagnetic package of track element 5 fixed at the track structure 7. Three-phase AC current is supplied to the anchor winding 4 generated magnetic field. Ferromagnetic packages of track element 5, actuated by excitation flux, interact with the field of traction winding, therefore it causes to traction effort. Magnetic attraction of the inductor to the track elements 5 creates the lifting effort. In case of transverse (lateral) deviation of the inductor under the effect of magnetic flux it is created the reaction transverse force, increasing in emergency operation modes when current supplying to the winding 3 of transverse stabilization of controlled DC.

For investigation the system with integral traction-levitation module it was manufactured the real prototype of vehicle bogie weighting 10 tons equipped with 10 LSM of 40 kW each [4].

In spite of the fact that it was succeeded to integrate functions of suspension and traction as well as the guidance system inside the single power element, as a whole the system is rather complicated, since each subsystem of traction, suspension and stabilization require the power supply and control systems.

Unfortunately, at this stage of traction levitation system development the research in Russia has been frozen. Renewed interest to magnetic levitation transport technologies causes to actuality of assessment of traction levitation systems' development potential in the context of their simplification.

3.   DEVELOPMENT POTENTIAL ASSESSMENTS OF TRACTION LEVITATION SYSTEMS

The further simplification of traction levitation system can be related with application of linear switched reluctance motor. The possibility to apply this type of electric machines is defined by the great value of normal force component between stator and rotor which can be used for generation of levitation and insurance of guidance system. Such kind of electric machine obtains the passive rotor consisted of ferromagnetic elements located along track structure. Rotors elements have great mechanical strength, which eliminates restrictions for transmission of mechanical traction force and suspension and gives the possibility to create the passive discrete track structure with reduced materials consumption, and at the same time the design of stator winding with concentrated coils is extremely simple.

In the range of switched reluctance machines there are two types which differ in direction of flux closure generated by motor's phase: with longitudinal flux design (Figure 4) coincided with moving direction, and with traverse design, vertical to moving direction (Figure 5).

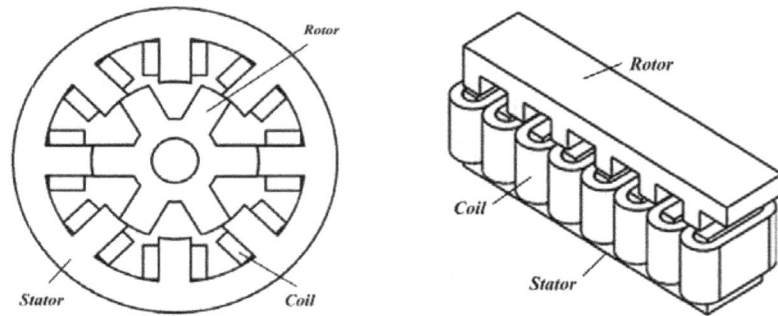

Figure 4. Switched reluctance motor with longitudinal flux design

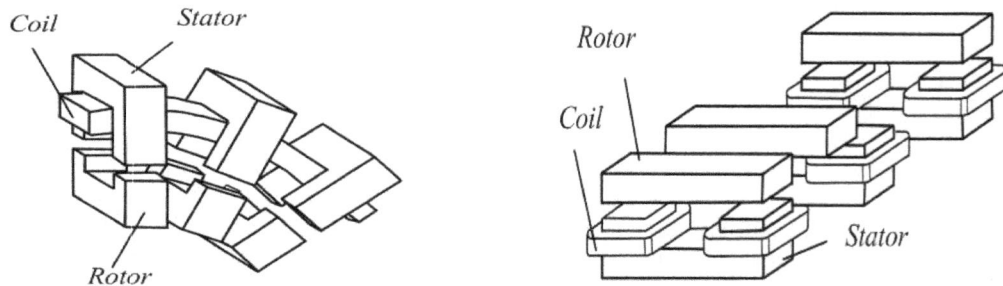

Figure 5. Switched reluctance motor with transverse flux design

These two linear configurations are the analogue of rotating machines and can be achieved as the result of stator and rotor turning about. These machines are simple, technologically efficient, mechanically enduring and have low loss; their power supply systems have simple circuits.

It is known the application of switched reluctance machines with longitudinal flux design in the traction levitation system of industrial conveyors [5].

The design of traction levitation system is proposed where one group of linear machines ensures levitation and traction, the second one ensures traction and guidance system. Therefore, the necessity to develop the separate traction, suspension and stabilization systems eliminates. The proposed concept provides the transition from three different systems to two single-type systems where each of them is involved in traction setting and as a whole it leads to cost reduction.

Mechanical configuration of vehicle traction levitation system developed by this concept can be presented at Figure 6. Rotors 3 and 5 of linear switched reluctance motor with longitudinal flux design are installed at the track structure 1 in its lower horizontal and side edges; motor's stator 4 and 6 are installed horizontally and vertically at the vehicle bogie frame. Horizontally installed motors generate traction and lifting effort, and vertically installed motors in addition to traction effort in running direction generate the force in the horizontal plane, vertical to running direction. The guidance system is provided by adjusting the ratio of forces values of the left and right vehicle sides.

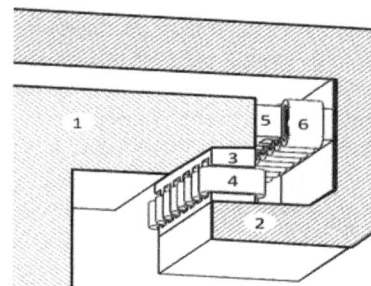

Figure 6. Concept scheme of traction levitation system

Simpler traction levitation system is possible to implement on the basis of linear switched reluctance motor with longitudinal flux. It can be used in two structural schemes. In the first scheme (Figure 7) the magnetic conductor of stator 1 with phase windings 2 are installed at the horizontal plane of bogie frame and the magnetic conductor of stator 4 is mounted at the track structure 5. U-shaped configuration of motor's magnetic conductor ensures the self- guidance system of the vehicle; the control is provided by two coordinates of traction and suspension. The disadvantage of this configuration is the mutual influence of traction and suspension as well as the weak passive guidance system which makes difficult guiding the vehicle in curves.

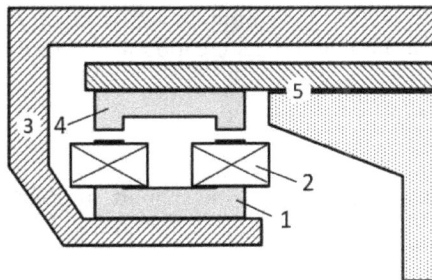

Figure 7. Mechanical configuration of traction levitation system

In the second scheme (Figure 8) U-shaped track is used. On its side edges discretely located passive elements of rotor 2 are mounted; magnetic conductors of stator 4 with phase windings 5 are installed on the sidewalls of a bogie frame 3.

The lifting force is generated when there is current supply in the phase windings and there is a displacement of teeth axles of magnetic conductors of stator 4 and rotor 2 relative to each other in the vertical plane. The width of vehicle vertical lifting can be the half of width of motor's tooth equaled to 25-50 mm. The change of phase current value of the motor does not have a great impact on levitation. Guidance system is provided by adjusting the ratio of total forces generated in horizontal planes by motors of each bogie side. In considered configuration the traction windings merge the functions of traction setting, levitation and guidance system; they are powered by a single converter which greatly simplifies the system and improves its energy indicators.

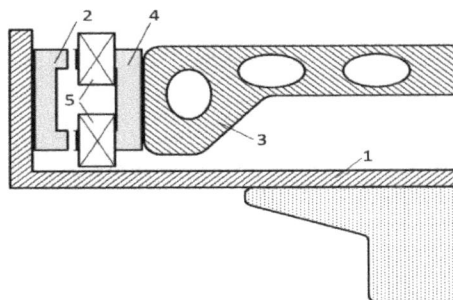

Figure 8. Mechanical configuration of traction levitation system

4.   THE FORECAST OF TRACTION LEVITATION SYSTEMS DEVELOPMENT

The most important focus area in technology improvement is to forecast its development allowing us to formulate the directions for further research. In the context of the topic it can be assumed that having reached the limit of simplicity, the traction levitation system with the switched reluctance motor will predetermine the beginning of a new round of its sophistication, associated with natural tendency to reduce the working clearance of electric machine, and it will cause to requirements for improvement the dynamic properties of the system and to necessity to apply new magnetic materials.

5. CONCLUSION

1) During development of traction levitation system of vehicle with magnetic suspension there is a tendency to simplify the mechanical configuration. The tangible embodiment was received by idea to integrate different functions of traction levitation and lateral stabilization in a single power element designed on the basis of linear reluctance motor.

2) Development potential of traction levitation systems in terms of their simplification relates with application of linear switched reluctance motor with longitudinal flux. The system simplification can be reached by merging the functions of levitation, traction and guidance system in the single power element as well as by management through the one current channel of stator winding.

3) It is possible to predict the possibility of development the simple traction levitation system for high-speed vehicles with magnetic suspension and the possibility to apply during development well studied technical means, which significantly reduce the cost for its creation and contribute to its wide implementation in commercial operation.

ACKNOWLEDGEMENTS

The presented work has been developed with support of Russian Ministry of Education, grant 14.576.21.0040.

REFERENCES

[1]   Antonov YF. *Magnetic Levitation Technology: Scientific Problems And Technical Solutions*. Magnetic levitation transport systems and technologies: materials of the 1st International Scientific conference, St. Petersburg, 2013.

[2]   YA Bakhvalov, VA Vinokurov. Vehicle with magnetic suspension, *Machine building*, 1991; 320.

[3]   VM Pavlukov, LF Kolomeytzev, FA Rednov, Linear synchronous electric motor, 1992; 29.

[4]   Talia II. System of linear electric drive and magnetic suspension based on linear reluctance electric motor for the coach of urban and suburban service, *Electric locomotive*, Novocherkassk, 1990; 31: 175-187.

[5]   Ramu et al, Transportation System with Linear Switched Reluctance Actuator for Propulsion and Levitation, 2004.

15

Optimal Planning of an Off-grid Electricity Generation with Renewable Energy Resources using the HOMER Software

Hossein Shahinzadeh, Gevork B. Gharehpetian, S. Hamid Fathi, Sayed Mohsen Nasr-Azadani
Department of Electrical Engineeing, Amirkabir University of Technology, Tehran, Iran

ABSTRACT

Keyword:

Batteries
Diesel Generators
HOMER Software
PhotovoltaicSystems
Renewable EnergyResources

In recent years, several factors such as environmental pollution which is caused by fossil fuels and various diseases caused by them from one hand and concerns about the dwindling fossil fuels and price fluctuation of the products and resulting effects of these fluctuations in the economy from other hand has led most countries to seek alternative energy sources for fossil fuel supplies. Such a way that in 2006, about 18% of the consumed energy of the world is obtained through renewable energies. Iran is among the countries that are geographically located in hot and dry areas and has the most sun exposure in different months of the year. Except in the coasts of Caspian Sea, the percentage of sunny days throughout the year is between 63 to 98 percent in Iran. On the other hand, there are dispersed and remote areas and loads far from national grid which is impossible to provide electrical energy for them through transmission from national grid, therefore, for such cases the renewable energy technologies could be used to solve the problem and provide the energy. In this paper, technical and economic feasibility for the use of renewable energies for independent systems of the grid for a dispersed load in the area on the outskirts of Isfahan (Sepahan) with the maximum energy consumption of 3Kwh in a day is studied and presented. In addition, the HOMER simulation software is used as the optimization tool.

Corresponding Author:

HosseinShahinzadeh,
Department of Electrical Engineeing,
Amirkabir University of Technology (Tehran Polytechnic),
Tehran, Iran.
Email: H.S.Shahinzadeh@ieee.org, Shahinzadeh@aut.ac.ir.

1. INTRODUCTION

With the world's industrial development and the growing demand for energy and on the other hand limited fossil fuels and the necessity of preserving resources for future generations and also preventing environmental damage caused by burning them, there is no other way left but to use renewable energies.

Absence of national grid in remote areas, high cost of construction of a new transmission lines due to the long distances and geographical complications, made the administrators and designers of the electricity grid to think about looking for alternative solutions to supply energy in such areas. On the other hand, the increasing rate of electrical energy and large number of dispersed consumers has become one of the biggest problems for power companies. This factor made the power companies to think of using renewable energy as a solution to provide power for the independent loads of the grid.

One of the most important renewable energies is solar energy which is as a free and inexhaustible energy source, has the potential to transform into other forms of energy and in the form of photovoltaic systems can be used as a cost effective source of electrical energy to provide electrical energy for the consumers that have no access to the national transmission grid due to geographical and climatic conditions.

Using diesel generators for many years was considered as the best solution for power supplying dispersed loads which are far from grid. But nowadays, due to the advancements in renewable energies' technology and environmental concerns, energy production by renewable energy sources is expanded more and more.

Today, hybrid systems have become one of the most effective solutions to meet the electrical energy needs of different regions. Having in mind the discontinuation of produced energy by the renewable sources, in practice it has been proven that the use of hybrid systems can be an appropriate solution and using a proper combination of these resources, an affordable, clean and reliable production system is attainable.

Iran that its latitude is 25 to 45 degree is one of the ideal areas in terms of solar radiation. The minimum annual average of solar radiation on the horizon is approximately 3.6 $\frac{Kwh}{m^2day}$ for Rasht and the maximum of it is about 5.9 $\frac{Kwh}{m^2day}$ for Bam. In the sloped surface (with 45 degree), the average amount of the solar radiation is 6.5 $\frac{Kwh}{m^2day}$. In Iran, except the Caspian Sea coasts, the percentage of sunny days throughout the year is between 63 to 98 percent. That means more than 300 days of the year is sunny. Considering this amount of solar radiation, it can be said that most provinces are located in proper solar radiation area [1].

The purpose of this study is the optimization and technical and economic analysis of the hybrid systems for a load far from grid in an area on the outskirts of Isfahan (Sepahanshahr). Studies had done using solar radiation data in the area and the cost of diesel fuel and the output parameters are expressed as functions of these variables.

In this study HOMER software is used for simulation. Homer has been produced and extended by the International Organization of Renewable Energies. This software could be used for sizing hybrid systems which is based on the net present costs. In addition, this software is able to perform sensitivity analysis on variables with non-deterministic values. In fact, HOMER makes it possible for the user to check the effect of changing a variable over the entire system. This software requires data on energy sources such as: type of system components, number of components, costs, efficiency, longevity, economic constraints, and control methods for the analysis [2].

In this paper, at first the conditions and properties of the site is expressed and then structure of the proposed hybrid system for providing load energy in case study in Sepahanshahr is studied in detail and this proposed system is modeled in HOMER. Then, more detailed information on the various components of the system such as load, resources etc. are presented. Finally, the results of the HOMER software simulation are presented.

2. EVALUATION OF CONDITIONS OF STUDY SITE

Isfahan is located in 435 kilometers south of Tehran in the central plateau of Iran. General level of Isfahan is about 1570m above sea level. Isfahan with the longitude of 51 degrees 39 minutes 40 seconds east and latitude of 32 degrees 38 minutes 30 seconds north have the average annual solar radiation of about 4.6 $\frac{Kwh}{m^2day}$ which indicates the considerable potential of solar energy in Isfahan. In this study, technical and economic evaluation of feasibility of installing and designing a hybrid system for providing electrical energy for a building with cultural and entertainment use on the district on the outskirts of Isfahan (Sepahanshahr) has been addressed. The building has two floors with beneficial underpin of 1350 square meters and total area of 1100 square meters. The extent and characteristics of the building's electrical load requirements will be provided in the next sections. Figure 1 depicts the exact location of the building and site which is marked on aerial photograph [3].

Figure 1. Location of the studied building marked on the aerial photograph

3. HYBRID SYSTEM

A hybrid system is consisted of two or more sources of power generation which is used to obtain higher efficiency than systems which include only one source of power and exploitation in the best work point. Since the operation of a hybrid power generation system strongly depends on the environmental conditions of its usage, so it seems necessary to firstly choose the renewable energy sources which are appropriate to the potential of the area of exploitation of this hybrid system. The problem of discontinuity using renewable sources in hybrid systems could be eliminated by using energy storage elements. Figure 2 shows the overall schematic of the energy production system independent of grid which is considered in this study [4].

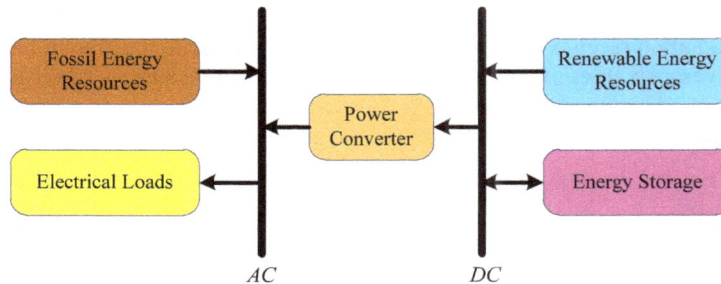

Figure 2. Overall schematic of the energy production system independent of grid in this study

The overall schematic shown in Figure 2 is modeled in HOMER software like Figure 3. Details on the various components of the system such as load, sources etc. are provided in the next sections.

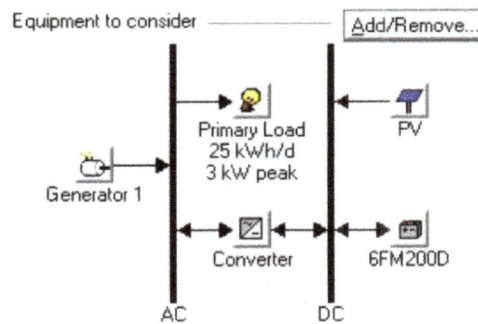

Figure 3. Implementation of production system independent of grid in the HOMER software

4. THE PARAMETERS AND CONDITIONS OF THE STUDY SITE
4.1. Electrical Load

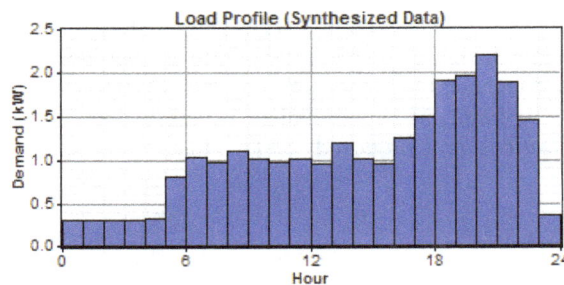

Figure 4. Daily load profile of the studied complex in a particular day

According to the calculations which were made, in the studied cultural and entertainment complex in Sepahanshahr area which is considered as an independent load, the amount of the consumed electrical energy is 25Kwh/d and the maximum demand is 3kw. According to calculations and measurements which were made on a particular day, the maximum peak load in the building was 2.33kw and the maximum peak load in this complex was between the hours of 17 to 22. Figure 4 depicts the daily load profile of this cultural and entertainment complex in a particular day [5].

4.2. Solar Power

As stated the study site is located in Sepahanshahr area in suburbs of Isfahan. With annual average of solar radiation of about $4.6 \frac{Kwh}{m^2 day}$, Isfahan has an appropriate potential to use solar energy. Table 1 shows the average values of radiation received in a horizontal surface of the earth in different months of the year.

By entering Average values of radiation received in a horizontal surface of the earth in HOMER and considering the height of the studied site, an index introduced calling clearness index. Figure 5 shows the output related to the solar radiation in different months in Isfahan [6].

Table 1. Average Values of Radiation Received in a Horizontal Surface of the Earth in Different Months of the Year in Isfahan

Month	Average radiation $\frac{Kwh}{m^2 day}$
January	2.694
February	3.444
March	4.083
April	4.972
May	5.889
June	6.638
July	6.305
August	5.833
September	5.223
October	4.027
November	2.944
December	2.583
Average	4.552

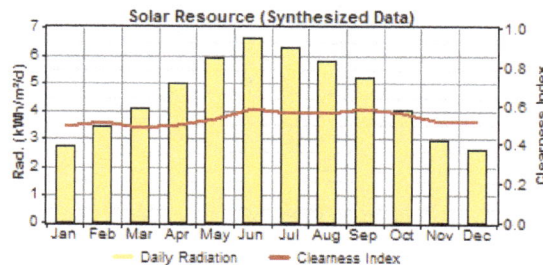

Figure 5. Amount of the radiation in Isfahan ($\frac{Kwh}{m^2 day}$)

5. THE PARAMETERS AND CONDITIONS OF THE STUDY SITE

In this section we review the components of the hybrid system which is used in this case study and we will try to provide proposed technical and economic models. Components of the hybrid system used in this case study are: 1) Diesel Generator, 2) Photovoltaic, 3) Battery, and 4) Converter. In the following we describe the technical specifications and proposed model along with the price, hours of operation and characteristic of each unit.

5.1. Diesel Generator

One of the most important parts of designing hybrid systems for providing electrical energy is the proper selection of a diesel generator, because if it is not selected properly, irreparable damages could occur. The most significant parameters to consider when selecting a diesel generator are: 1) Whether diesel generators are used to supply emergency power or it is going to be used permanently. 2) Type of the load that diesel generator is used to provide. 3) KW and KVA of the diesel generator.

Unfortunately most of the people believe that low power diesel generator is more suitable for emergency power usage, because these types of diesel generators are used in case of power failures on part-time. This misunderstood of the consumers often results in damages to diesel generator and the devices which are connected to it. Hence one should consider the load amount in case of power failures and thenselect the needed diesel generator [7].

Cost of the commercial diesel generators on the market varies depends on the size of the unit and their power capacity. For this study, the purchase price (based on the prices offered by Tabriz Motorsazan Company) varies between 124 to 140 dollars per kilowatts. Therefore in this analysis the cost of purchasing and installing is considered 160 dollars per kilowatt and the cost of replacement and maintenance are considered 143 and 0.08 dollars per kilowatt accordingly. In figure 6 the curve of capacity and its costs are shown [8].

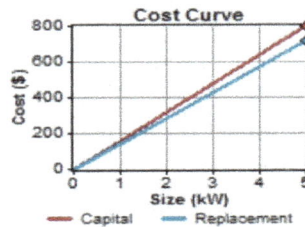

Figure 6. Curve of costs of installation and replacement of diesel generator in HOMER

5.2. Photovoltaic System

Solar panels selected for this study are solar panel 250W with crystal menu of LG Company (LG250s1c), that the technical specifications of the panel are presented in Table 2 and 3.

Table 2. Electrical Characteristics at Standard Test Conditions(STC)

Maximum Power at Standard Test Conditions (P_{max})	250
Voltage at the Maximum Power Point(V_{mpp})	29.9
Current at the Maximum Power Point(I_{mpp})	8.37
Open Circuit Voltage (V_{oc})	37.1
Short Circuit Current (I_{sc})	8.76
Module Efficiency (%)	15.5
Operation Temperature (°C)	-40 °C ~ +90 °C
Maximum System Voltage (V)	1000
Maximum Series Fuse Rating (A)	15
Power Tolerance	0 ~ +3 %

Standard Test Conditions (STC): Radiation Rate 1000 W/m², Temperature 25°C, Time 1.5AM.

Table 3. Electrical Characteristics of Normal Operation Conditions and Temperature(NOCT)

Maximum power (W)	176
Maximum power voltage (V)	27.35
Maximum power current (A)	6.42
Open circuit voltage (V_{oc})	34.54
Short circuit current(I_{sc})	6.77
Efficiency reduction (from 1000 W/m² to 200 W/m²)	< 4.5 %

Normal Operation Conditions and Temperatures : Radiation Rate 800 W/m2, Temperature 20°C, Wind Speed 1m/s.

The purchase cost of solar panels (LG250s1c) is about 1.6 dollars per watt. Hence in this study, the purchase and installing cost of solar panels is considered 2000 dollars per kilowatt and the replacement cost of these panels is considered 1700 dollars. Figure 7 shows the curve of capacity and costs of solar panels [9].

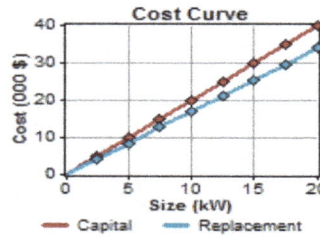

Figure 7. Curve of the costs of installation and replacement of solar panels in HOMER

In this study, photovoltaic system in sizes of 0, 2.5, 5, 7.5, 10, 12.5, 15 kW is used. Intended lifespan for this solar system is 20 years.

5.3. Battery
Different models of batteries are available in the market for this purpose. The selected battery for this study is 6FM200D (Ah 200, V12) from Vision Company, that its technical specifications is given in Table 4 and the discharge curve is presented in Figure 8.

Table 4. Technical Specification of Battery (6FM200D)

Nominal Voltage (V)	12
Number of Cells	6
Designed Lifetime (year)	10
Nominal Operational Temperature($°C$)	25 °C
Internal Resistance of the Battery at Full Charge($mOhms$)	3.5
Efficiency of Module (%)	15.5
Storage Temprature($°C$)	-20 °C ~ +60 °C
The Maximum Discharge Current at 25 °C (A)	1000 A ($5s$)
Short Circuit Current (A)	3300
Maximum Charge Current(A)	60

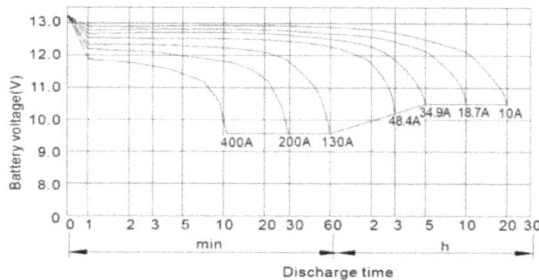

Figure 8. Battery discharge curve (6FM200D)

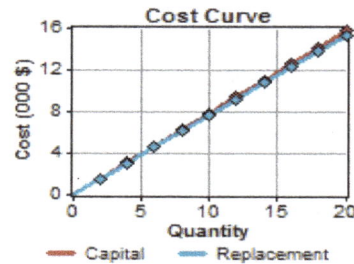

Figure 9. Curve of costs of installing and replacing batteries in HOMER

The purchase cost of this model of battery (6FM200D) is about 755 dollars for each battery. So in this study the cost of purchasing and installing each battery is considered 790 dollars and replacement cost is considered 770 dollars for each battery. Figure 9 shows the curve of number and cost of batteries [10].

In this study, the number of batteries in system is used as 0, 4, 8, 12, 16, 20, 24, 30. Intended lifetime for each battery is 10 years.

5.3. Converters
Purchase cost of a DC to AC converter (based on the prices offered by Sunrous Company) is about 180 dollars per kilowatt. So in this study, the purchase and installation cost of converter is considered 200 dollars per kilowatt and replacement cost is considered about 180 dollars. In this study, converters with the size of 0, 2, 4, 6, 8 kilowatt are used in system. Lifetime of the unit is 10 years and its efficiency is 98%. In figure 10 curves of capacity and its cost is displayed [11], [12].

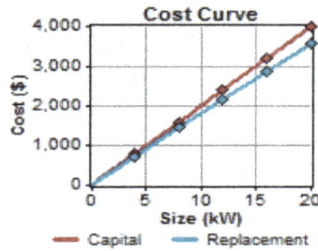

Figure 10. Curve of the costs of installation and replacement of converter in HOMER

6. RESULTS OF THE SIMULATION PERFORMED IN HOMER SOFTWARE

In HOMER software for calculation of system lifetime, Net Price Calculation (NPC) equation is used, in which the costs include installation, replacement, fuel, etc. all costs and revenues are assessed with a fixed interest rate during the year. In order to considering impact of inflation in the calculations, Equation (1) is used.

$$i = \frac{i'-f}{1+f}$$
(1)

In this,

i: Real interest rate,

i' : Nominal interest rate,

f: The inflation rate.

The main output of the economic calculations in this software is Net Price Calculator (NPC) which is calculated from the Equation (2):

$$C_{NPC} = \frac{C_{annual,\ total}}{CRF\left(i, R_{Project}\right)}$$
(2)

In it,

$C_{annual,\ total}$: Total annual cost,

$R_{Project}$: Lifetime of the project,

i : real interest rate.

To calculate the return on capital over the N years, Equation (3) is used:

$$CRF\left(i, N\right) = \frac{i\left(1+i\right)^{N}}{\left(1+i\right)^{N}-1}$$
(3)

In which $CRF\ (I,N)$ is the return on capital factor during N years.

In the optimization which is done in HOMER, all of the possible states have been simulated and the best combination with the minimum (NPC) is introduced as the optimum arrangement [13]. This best combination provides the all predetermined constraints set by the operator with the lowest net cost. In this paper, in addition to the constraints about fuel, cost, etc. the constraint of minimum penetration of renewable sources is added too. The result of optimization by HOMER is shown in Figure 11. In the most optimized condition that is considered, photovoltaic system and storage have the capability of supplying all of the required electrical power for the load and the diesel generator will play the role of a resource in the system.

Figure 11. Results of the Optimization Problems in HOMER

Details of the various costs of the components of the studied hybrid system in 20-year lifetime of the project are shown in Figure 12. As it can be seen in Figure 12, the maximum initial cost in the project is about purchasing photovoltaic system but the maximum replacement cost during 20 year is for battery which has a great role in the overall cost of the project.

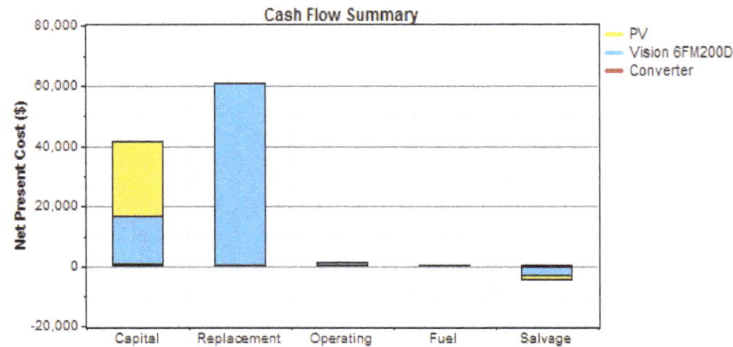

Figure 12. Cost of different components of the studied hybrid system

Accordingly, the average power produced by the solar cells is 20.335 (kWh/yr), but this rate varies in different months of the year. The maximum power produces by the solar cells is in June and the minimum of it is for December. In Figure 13 the average power produced by the solar cells for different months of the year are shown.

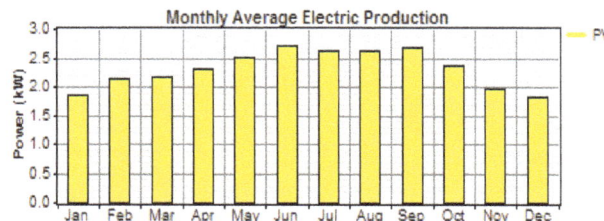

Figure 13. The average power produced by the solar cells for different months of the year

Accordingly the output power of the solar system in different hours of the day based on the capacity and produced power for different months of the year is displayed in Figure 14.

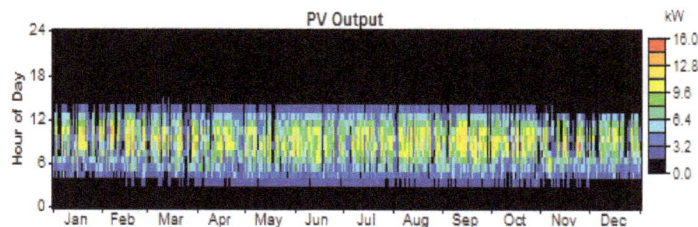

Figure 14. The electric power output of the solar system in different hours of the day

Excess electrical power generated by the solar system is stored in the considered storage system, batteries. The rate of the stored electrical power in the batteries is shown for different hours of day based on the capacity and produced power for different months of the year in Figure 15.

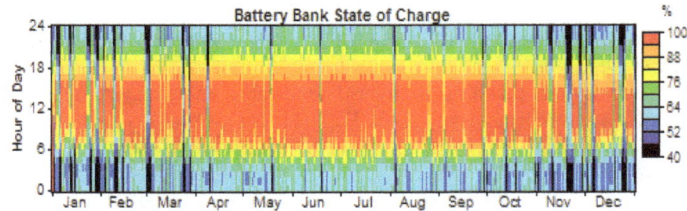

Figure 15. The amount of power stored in the batteries for different hours of day

To further explore the behavior of the designed hybrid system in providing the required electrical energy for the load, the generated electric power by solar cells, stored and discharged energy by the batteries for a week is selected and depicted in Figure 16. As it can be seen in the figure, the considered solar system and the storage for this study can jointly provide the required electrical load of the system entirely. In this system, in the hours that the generated energy by the solar cells is greater than the required electrical power, this excess is stored in the intended batteries and in the hours as the night that solar cells cannot provide electrical power, batteries go into the circuit and provides the required electrical energy of the load. Thus in the designed hybrid system the all required electrical power for the load is provided by the photovoltaic and storage system [14], [15].

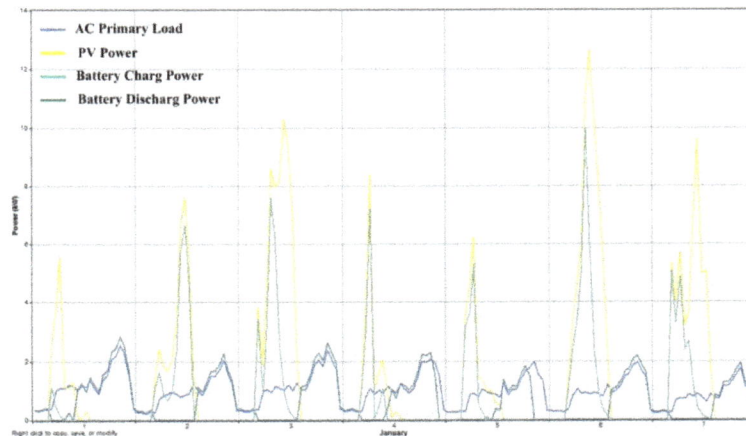

Figure 16. Evaluation of the studied hybrid system in a week

7. CONCLUSION

With industrial development of the world and the increasing demand of energy in one hand and limited resources of fossil fuels and the need to conserve these resources for the next generations and also preventing environmental damage caused by burning them on the other hand, there is no other solution than using renewable energies such as solar. Furthermore, around the world one of the major concerns about remote areas and areas far from the grid is providing electrical power for these areas. Connecting these areas to the grid is costly and in some cases is physically impossible. In such case using Distributed Generation (DG) resources id the best option for providing required electrical energy for this category of consumers. In the coming years with reduction in initial investment cost as well as increasing the efficiency of solar panels, the importance and necessity of using these panels to provide clean energy appears more than even before. In addition, with producing energy near the consumption centers eliminates the need to establish voltage transmission lines near towns and villages. Since the hybrid systems are fed from two or more sources of energy, they have higher reliability in compare with systems that have only a source for energy generation. In this paper, a hybrid system was proposed to provide the electrical energy for the studied load. Using the HOMER software optimizations of the system was performed and the optimal mode was selected. In this case 100% of the required energy is supplied by the intended photovoltaic and storage system and the diesel generator is a backup source.

REFERENCES

[1] Bahrami, Mohsen, Payam Abbaszadeh. An overview of renewable energies in Iran. *Renewable and Sustainable Energy Reviews*. 2013; 24: 198-208.

[2] Fulzele JB, SubrotoDutt. Optimium planning of hybrid renewable energy system using HOMER. *International Journal of Electrical and Computer Engineering (IJECE)*. 2011; 2(1): 68-74.

[3] Alamdari, Pouria, Omid Nematollahi, Ali Akbar Alemrajabi. Solar energy potentials in Iran: A review. *Renewable and Sustainable Energy Reviews*. 2013; 21: 778-788.

[4] Shaahid SM, LM Al-Hadhrami, MK Rahman. Review of economic assessment of hybrid photovoltaic-diesel-battery power systems for residential loads for different provinces of Saudi Arabia. *Renewable and Sustainable Energy Reviews*. 2014; 31: 174-181.

[5] Asrari, Arash, Abolfazl Ghasemi, Mohammad Hossein Javidi. Economic evaluation of hybrid renewable energy systems for rural electrification in Iran—A case study. *Renewable and Sustainable Energy Reviews*. 2012; 16(5): 3123-3130.

[6] Mohajer, Alireza, Omid Nematollahi, Mahmood Mastani Joybari, Seyed Ahmad Hashemi, Mohammad Reza Assari. Experimental investigation of a Hybrid Solar Drier and Water Heater System. *Energy Conversion and Management*. 2013; 76: 935-944.

[7] Shiroudi, Abolfazl, Seyed Reza Hosseini Taklimi, Seyed Ahmad Mousavifar, Peyman Taghipour. Stand-alone PV-hydrogen energy system in Taleghan-Iran using HOMER software: optimization and techno-economic analysis. *Environment, development and sustainability*. 2013; 15(5): 1389-1402.

[8] Khatib, Tamer, A Mohamed, K Sopian, M Mahmoud. Optimal sizing of building integrated hybrid PV/diesel generator system for zero load rejection for Malaysia. *Energy and Buildings* 2011; 43(12): 3430-3435.

[9] Diaf, Said, Djamila Diaf, Mayouf Belhamel, Mourad Haddadi, Alain Louche. A methodology for optimal sizing of autonomous hybrid PV/wind system. *Energy Policy*. 2007; 35(11): 5708-5718.

[10] Fulzele JB, Subroto Dutt. Optimium planning of hybrid renewable energy system using HOMER. *International Journal of Electrical and Computer Engineering (IJECE)*. 2011; 2(1): 68-74.

[11] Shahinzadeh, Hossein, Mohammad Moien Najaf Abadi, Mohammad Hajahmadi, Ali Paknejad. Design and Economic Study for Use the Photovoltaic Systems for Electricity Supply in Isfahan Museum Park. *International Journal of Power Electronics and Drive Systems (IJPEDS)*. 2013; 3(1): 83-94.

[12] Kaabeche, Abdelhamid, RachidIbtiouen. Techno-economic optimization of hybrid photovoltaic/wind/diesel/battery generation in a stand-alone power system. *Solar Energy*. 2014; 103: 171-182.

[13] Al-Karaghouli Ali, LL Kazmerski. Optimization and life-cycle cost of health clinic PV system for a rural area in southern Iraq using HOMER software. *Solar Energy*. 2010; 84(4): 710-714.

[14] Eltamaly, Ali M, et al. Economic Modeling of Hybrid Renewable Energy System: A Case Study in Saudi Arabia. *Arabian Journal for Science and Engineering*. 2014; 1-13.

[15] Sen, Rohit, Subhes C Bhattacharyya. Off-grid electricity generation with renewable energy technologies in India: An application of HOMER. *Renewable Energy*. 2014; 62: 388-398.

16

Estimation of Harmonics in Three-phase and Six-phase (Multi-phase) Load Circuits

Deepak Kumar, Zakir Husain

Departement of Electrical Engineering, National Institute of Technology, Hamirpur (HP), India

ABSTRACT

Keyword:

Fast Fourier Transform
Harmonics
Inverters
Multi phase system
Ripple
Total Harmonic Distortion

The Harmonics are very harmful within an electrical system and can have serious consequences such as reducing the life of apparatus, stress on cable and equipment etc. This paper cites extensive analytical study of harmonic characteristics of multiphase (six- phase) and three-phase system equipped with two & three level inverters for non-linear loads. Multilevel inverter has elevated voltage capability with voltage limited devices; low harmonic distortion; abridged switching losses. Multiphase technology also pays a promising role in harmonic reduction. Matlab simulation is carried out to compare the advantage of multi-phase over three phase systems equipped with two or three level inverters for non-linear load harmonic reduction.The extensive simulation results are presented based on case studies.

Corresponding Author:

Deepak Kumar,
Departement of Electrical Engineering,
National Institute of Technology,
Hamirpur, Himachal Pradesh, India-177005.
Email: dkaroraelectrical@gmail.com, timothyarora@gmail.com

1. INTRODUCTION

Owing to the budding benefits resulting from the use of a phase order higher than three in multi-phase transmission and distribution, some interest has also developed in the area of multi-phase system analysis in recent past [1]-[10]. Also, multi-level inverters have emerged as a capable tool in achieving high power ratings with voltage limited devices. This paper presents a functional model of two and three level inverter with multi and three phase load and simulation of the developed model is done with the help of MATLAB/Simulink. Multi-level inverter fed electric machine drive systems have emerged as a promising tool in achieving high power ratings with voltage limited devices [20]. The conventional inverters used are voltage source inverter (VSI) and current source inverter (CSI) which consists of a dc link and Inverter Bridge. Harmonic reduction is achieved to greater extent than conventional inverter such as voltage source inverter, current source inverter in multilevel inverter and multiphase loads. High phase number drives own several advantages over conventional three-phase drives such as: reducing the amplitude and increasing the frequency of torque pulsation, reducing the rotor harmonic currents, reducing the current per phase without increasing the voltage per phase, lowering the dc link current harmonics, higher reliability and increased power. Harmonics are very detrimental within an electrical system and can have serious consequences. For example, the presence of harmonics reduces the life of apparatus. Harmonics cause things to run hot, which cause stress on the cables and equipment. In the long term, this degrades an electrical system. The presence of harmonics will also mean that although you will get billed for the power that you are supplied, a large percentage of that power may be not viable. Harmonic mitigation is taking action to minimize the presence of harmonics in your electrical system and can achieve great cost savings. Harmonic distortion can cause poor power factor, transformer and distribution equipment overheating, random breaker tripping, or even sensitive equipment failure. Since harmonics affect the overall power distribution system, the power utility may even levy heavy fines when a facility is affecting the utilities' ability to efficiently supply power to all of its customers. These harmonics can be suppressed using multilevel inverter equipped with multi-phase loads.

The multi-phase technology received a substantial worldwide attention by the various R&D's and front-end industries in three very specific application areas, namely electric ship propulsion, traction (including electric and hybrid electric vehicles) and the concept of 'more-electric' aircraft. Irrespective of abundant advantageous multi-phase electric drives are limited to economically viable design, power converter configurations and closed control aspects. Multi-phase power systems can be used to cancel harmonic currents. For higher power rectifier circuits, even 12-phase power systems have been used for further harmonic current reduction. Six phase transmission lines are popular due to its increased power transfer capability by $\sqrt{3}$ times, maintaining the same conductor configuration, better efficiency, better voltage regulation, greater stability and greater reliability.

2.   MULTI-LEVEL INVERTER

The power electronics device which converts DC power to AC power at required output voltage and frequency level is known as inverter. Inverters can be broadly classified into two level inverter and multilevel inverter. Multilevel inverter as compared to two level inverters has advantages like minimum harmonic distortion and can operate on several voltage levels. A multi-stage inverter is being utilized for multipurpose applications, such as active power filters, static var compensators and machine drives for sinusoidal and trapezoidal current applications. The drawbacks are the isolated power supplies required for each one of the stages of the multiconverter and it's also lot harder to build, more expensive, harder to control in software.

Multilevel inverters are named after the level of voltages that can be obtained from them. For example a 2-level inverter can take values +V and –V and 3-level inverter can produce voltage levels of +V, 0 and –V where V is the voltage of dc supply.

For 2 level inverter, there are two levels for phase voltage and three levels for line voltage as shown in following figure of matlab.

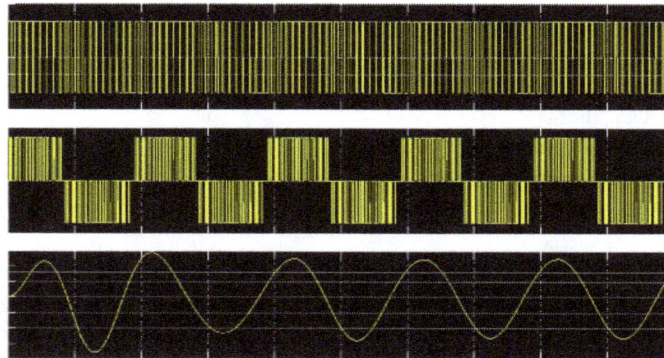

Figure 1. Van: inverter phase output voltage; Vab: Inverter line output voltage; Vab_load: inverter   load line voltage after linear transformer shown in the above figures respectively
(y axis: Voltage; x axis: Time)

The system consists of two independent circuits illustrating two three-phase two-level PWM voltage source inverters. Each inverter feeds an AC load through a three-phase transformer. Both converters are controlled in open loop with the Discrete PWM Generator block. The two circuits use the same DC voltage, carrier frequency, modulation index and generated frequency (f = 50 Hz). Harmonic filtering is performed by the transformer leakage inductance and load capacitance.

For 3 level inverter, there are three levels of phase voltage and 5 levels for line voltage as shown in following figure of matlab:

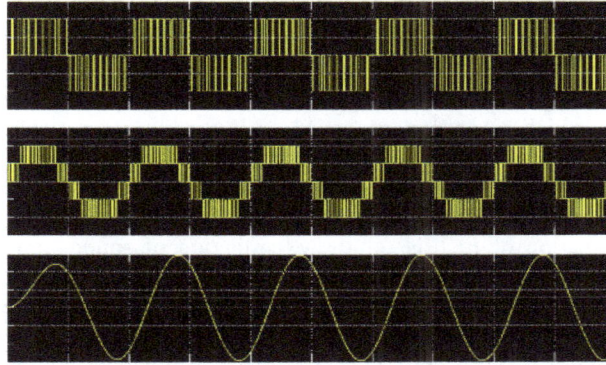

Figure 2. Van: inverter phase output voltage; Vab: Inverter line output voltage; Vab_load: inverter load line voltage after linear transformer shown in the above figures respectively. (y axis: Voltage; x axis: Time)

The system consists of two three-phase three-level PWM voltage source converters connected in twin configuration. The inverter feeds an AC load through a three-phase transformer. Harmonic filtering is performed by the transformer leakage inductance and load capacitance. Each of the two inverters uses the Three-Level Bridge block where the specified power electronic devices are IGBT/Diode pairs. Each arm consists of 4 IGBTs, 4 anti-parallel diodes, and 2 neutral clamping diodes. The inverter is controlled in open loop. Pulses are generated by the discrete 3-Phase Discrete PWM Generator block. This PWM generator or modulator can be used to generate pulses for 3-phase, 2-level, or 3-level converters using one bridge or two bridges The PWM modulator generates two sets of 12 pulses (1 set per inverter). The generator can operate either in synchronized or un-synchronized mode. When operating in synchronized mode, the carrier triangular signal is synchronized on a PLL reference angle connected to input 'wt'. In synchronized mode, the carrier chopping frequency is specified by the switching ratio as a multiple of the output frequency.

3. HARMONICS & FOURIER ANALYSIS

In three-phase power systems, even harmonics cancel out, so only the odd harmonics are of concern. On three-phase systems each phase voltage is 120 degrees out of phase, causing the phase current to be 120 degrees out of phase as well. With a sinusoidal voltage, current harmonics do not lead to average power. However, current harmonics do increase the rms current, and hence they decrease the power factor. The average power is:

$$\text{Pav} = \frac{V1 * I1}{2} \cos(\Phi 1 - \Phi 2) \qquad (1)$$

Where V1 and I1 are the peak values and, and $\Phi 1$ and $\Phi 2$ are the phase angles of fundamental voltage and current respectively. The rms current considering the harmonics is given by (2) as:

$$\text{Rms Current} = \frac{\sqrt{(I0^2 + \sum_{n=1}^{n=\infty} In^2)}}{2} \qquad (2)$$

Where I_n is the peak current at any harmonic number. With non-linear loads, the third harmonic on all three phases is exactly in phase and adds, rather than cancels, thus creating current and heat on the neutral conductor. Left un-treated, harmonic loads can reduce the distribution capacity and degrade the quality of the power of public utility power systems, increase power and AC costs, and result in equipment malfunctions such as communication errors and data loss.

A nonlinear load in a power system is characterized by the introduction of a switching action and consequently current interruptions. This behavior provides current with different components that are multiples of the fundamental frequency of the system. These components are called harmonics.

THD (Total harmonic distortion) is used as harmonic index for harmonic measurement which is given by:

$$THD = \frac{\sqrt{\sum_{n=2}^{n=\infty} In^2}}{I1} \tag{3}$$

THD is used as the harmonic index and harmonic spectrum is presented for each load using FFT (Fast Fourier Transform) [17], [18]. Fourier analysis of a periodic function refers to the extraction of the series of sines and cosines which when superimposed will reproduce the function. This analysis can be expressed as a Fourier series. The fast Fourier transform is a mathematical method for transforming a function of time into a function of frequency. Sometimes it is described as transforming from the time domain to the frequency domain. It is very useful for analysis of time-dependent phenomena.

One essential application of FFT is for the examination of sound. It is imperative to assess the frequency distribution of the power in a sound because the human ear exercises that capacity in the hearing process. For a sine wave with a single frequency, the FFT consists of a single peak. Combining two sound waves produces a complex pattern in the time domain, but the FFT clearly shows it as consisting almost entirely of two frequencies. For a full-wave rectified sine wave, meaning that the wave becomes positive wherever it would be negative. This creates a new wave with double the frequency. You can see that after rectification, the fundamental frequency is eliminated, and all the even harmonics are present.

Single-phase non-linear loads, like electronic ballasts, PC (Personal Computer) and other electronic apparatus, create odd harmonics (i.e. 3rd, 5th, 7th, 9th, etc.). Triplen harmonics (3rd order and its odd multiples) are bothersome for single phase loads because the A-phase triplen harmonics, B-phase triplen harmonics and C-phase triplen harmonics are all in the phase with each other. They will add rather than cancel on the neutral conductor of a 3-phase, 4-wire system. This can burden the neutral if it is not sized to handle this type of load. In addition, triplen harmonics cause circulating currents on the delta winding of a delta-wye transformer design. The result is transformer heating similar to that created by unbalanced 3-phase current. On the other hand, 3-phase non-linear loads like 3-phase ASDs, 3-phase DC drives, 3-phase rectifiers, etc., do not produce current triplen harmonics so much. These types of loads cause mainly 5th and 7th current harmonics and a minor amount of 11th, 13th, and higher order based on the design of the converter used.

4. THREE PHASE TO SIX PHASE TRANSFORMATION USING TRANSFORMER

Three phase voltages obtained from the inverter is fed to three single phase transformer for converting to six phase system. The circuit diagram for obtaining 6ϕ supply from 3ϕ supply using linear transformer is shown in Figure 3.

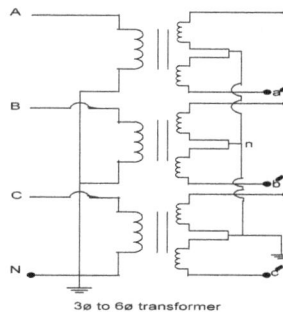

Figure 3. Linear transformer for three phase to six phase transformation

The transformer is called linear if the coils are wound on magnetically linear material (air, plastic, Bakelite, wood, etc.). Flux is proportional to current in the windings.

5. MATLAB MODELS

Four different cases are considered for the harmonic study as shown from Figure 4 to Figure 7 comprising of 2 and 3 level inverter with three phase or six phase rectifier circuit.

CASE I: Two level inverter- Measurement system- Three Phase Transformer- Three Phase Rectifier.

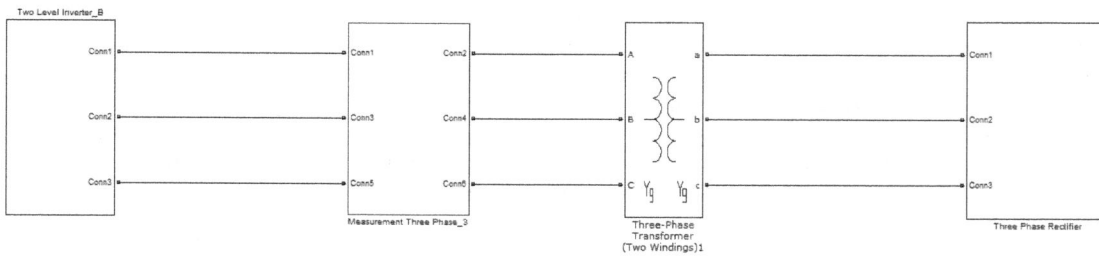

Figure 4. Two level inverter with three phase rectifier

CASE II: Two Level Inverter- Measurement System- Three Phase to six phase transformer- Six Phase Rectifier.

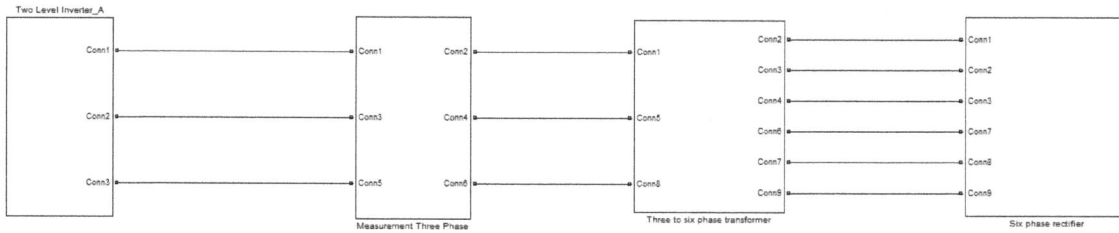

Figure 5. Two level inverter with six phase rectifier

CASE III: Three level inverter- Measurement system- Three Phase Transformer- Three Phase Rectifier.

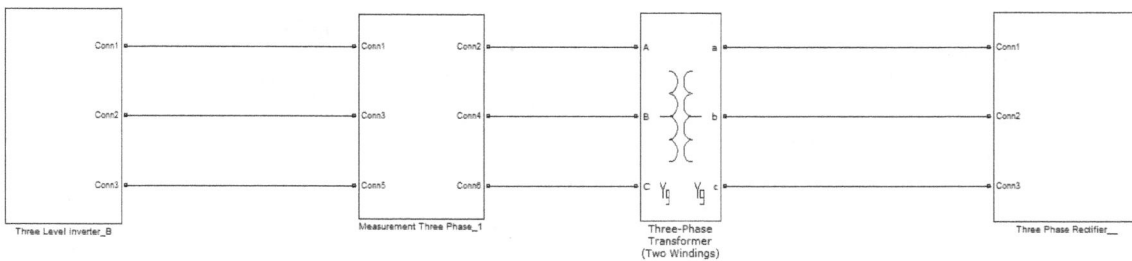

Figure 6. Three level inverter with three phase rectifier

CASE IV: Three level inverter- Measurement system- Three to six Phase Transformer- Six Phase Rectifier.

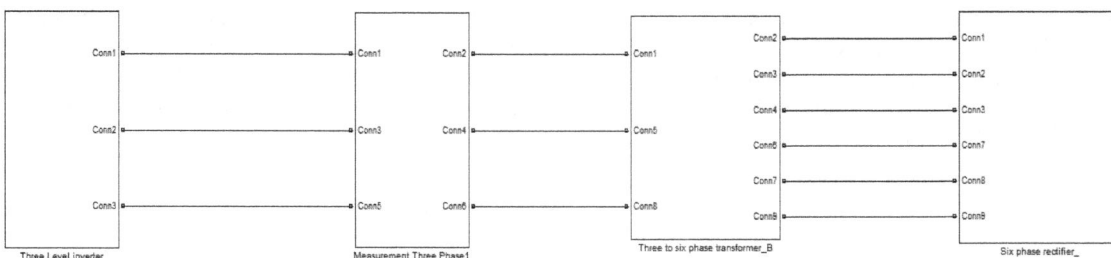

Figure 7. Three level inverter with six phase rectifier

Matlab simulink models of two and three level inverters are shown in Figure 8 and Figure 9 respectively. Three level inverter has advantage over two level inverter that when magnitude of supply is very high, use of filter is superfluous, constraint on the switches are low for the reason that the switching frequency may be low, and reactive power flow can be controlled. Two full-bridges VSI is employed wich contains twelve IGBT which switch on dc source [12]. Twelve pulses are generated for a double bridge three phase inverter. The first six pulses (1 to 6) fire the six devices of the first three arm bridge while the last six pulses (7 to 12) fire the six devices of the second three arm bridges.

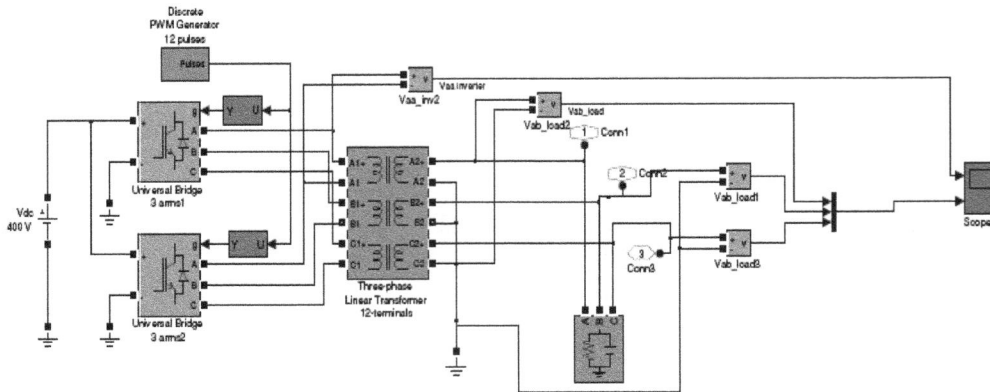

Figure 8. Matlab model of two level inverter

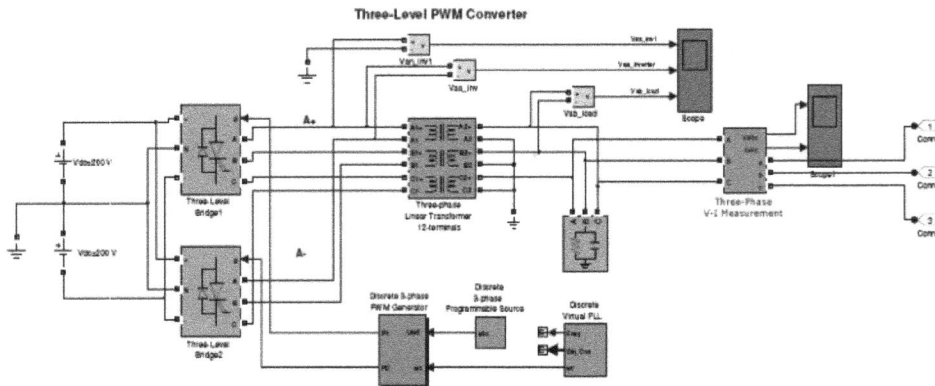

Figure 9. Matlab Model of three level inverter

A non-linear load on a power system is usually a rectifier and some kind of arc discharge device such as a fluorescent lamp, electric welding machine, or arc furnace in which current is not linearly related to the voltage. Current in these systems is interrupted by a switching action; the current contains frequency components that are multiples of the power system frequency and leads to distortion of the current waveform which in turn distorts the voltage waveform. For analysis of three phase non-linear load, RL load is replaced with three phase rectifier circuit. Following is the circuit of three phase rectifier. For six phase non-linear load, six RL load is replaced with two such circuits.

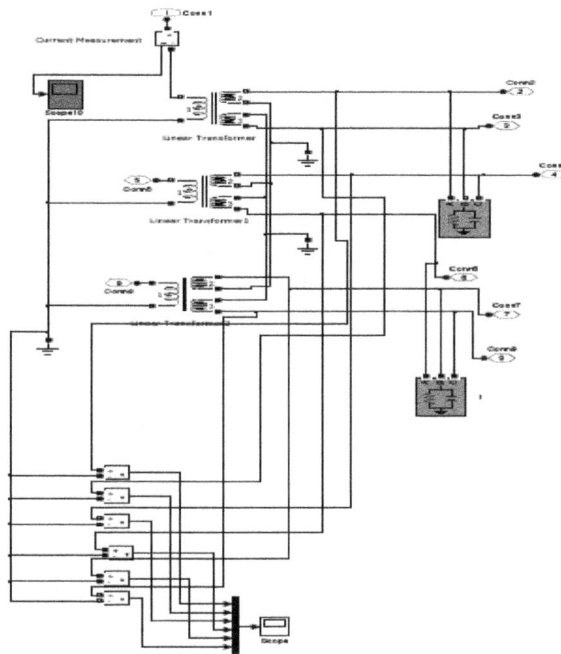

Figure 10. Matlab model of Three to six-phase transformer

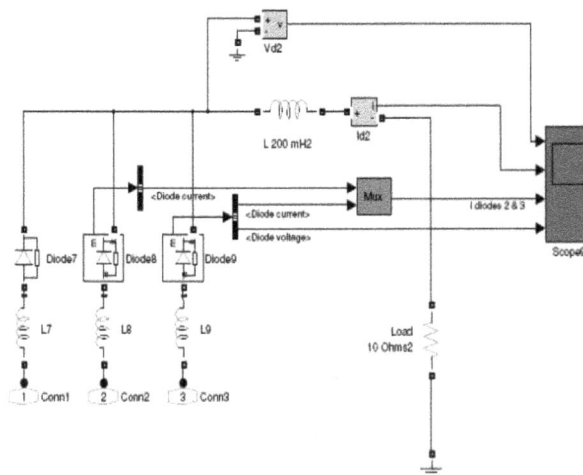

Figure 11. Three-phase rectifier circuit

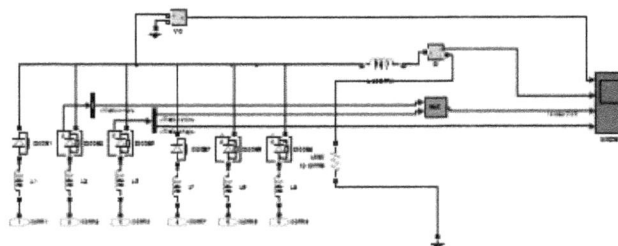

Figure 12. Matlab model of Six-Phase Rectifier

6. SIMULATION RESULTS

A nonlinear load in a power system is characterized by the beginning of a switching action and consequently current interruptions. This behavior provides current with different components that are multiples of the fundamental frequency of the system. These components are called harmonics which if not suppressed will cause severe problems in power distribution system. So, to analyze the effects of non-linear load, FFT analysis is done for the four different cases as mentioned above.

Once the simulation is completed, open the Powergui and select 'FFT Analysis' to display the frequency spectrum of signals saved in the structures. The FFT will be performed on a 2-cycle window starting at t = 0.1 - 2/50 (last 2 cycles of recording). Measurement of phase voltage FFT of load w.r.t ground for a nonlinear Load is shown from Figure 13 to Figure 16:

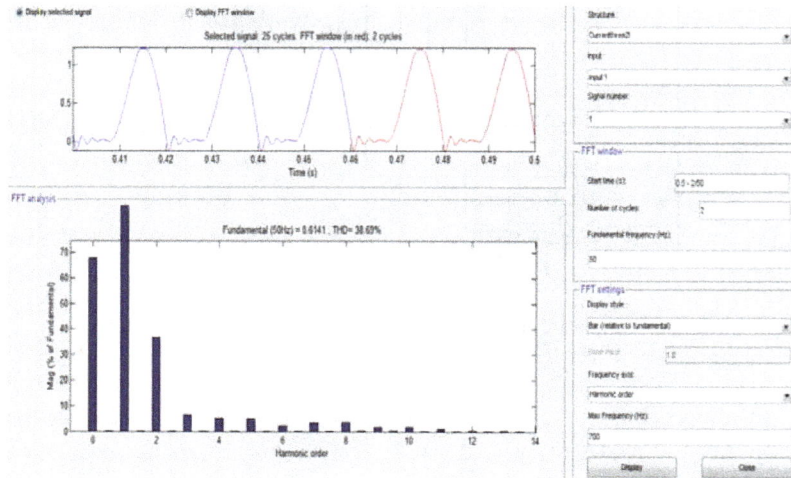

Figure 13. FFT analysis at two level inverter output with three phase rectifier load (Case I)

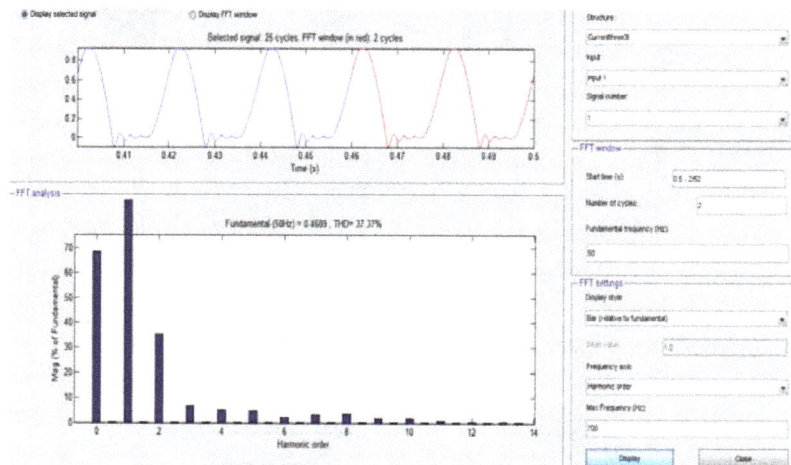

Figure 14. FFT analysis at three level inverter output with three phase rectifier load (Case III)

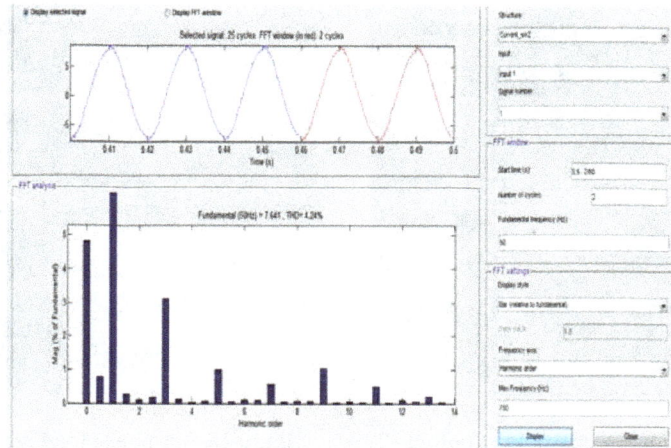

Figure 15. FFT analysis at two level inverter output with six phase rectifier load (Case II)

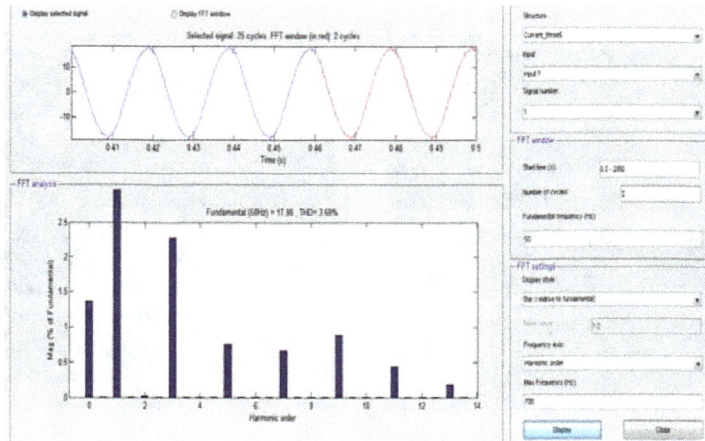

Figure 16. FFT analysis at three level inverter output with six phase rectifier load (Case-IV)

Total harmonic distortion (THD) for two level and three level inverter with three phase and six phase rectifier loads are shown in Table 1.

Type of inverter	Current THD for 3-phase rectifier load	Current THD for Multi-phase (6-phase) Rectifier load
Two Level Inverter	38.69%	4.24%
Three level inverter	37.37%	3.69%

For nonlinear current T.H.D (Total Harmonic Distortion) in the load current is found to be more in three phase. Power factor compensation requirement is more in three phase load as compared to three phase is also found by simulation. Rectified output voltage under these four different cases is also shown which shows that there are fewer ripples in six phase rectifier circuit.

From Figure 17 to 20, it is noted that ripple in case of six phase rectifier is less which is an undesirable factor in many electronic application as large ripples shorten the life of electrolytic capacitor, reduce the resolution of electronic test and measurement instruments etc.

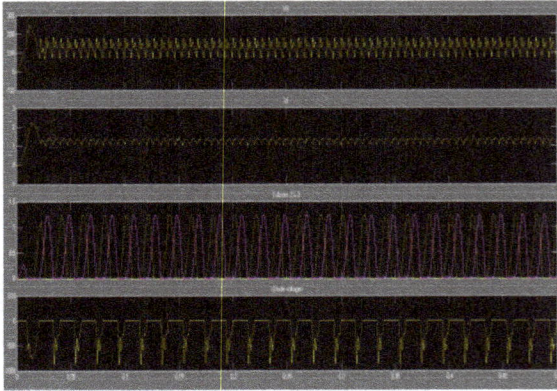

Figure 17. Three-phase load rectifier load output (2 level inverter): Y axis: Rectified output dc voltage, rectified output dc current, Current through diode, voltage across diode ; X axis: time. (Case I)

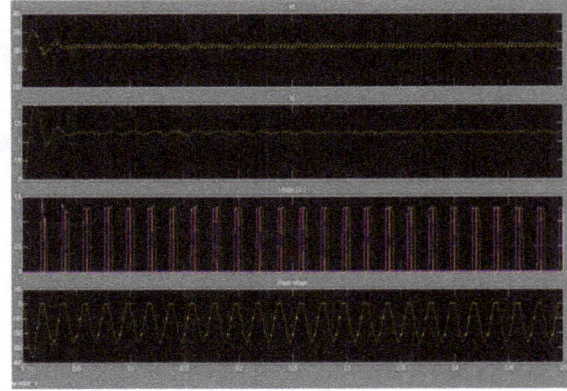

Figure 18 Six-phase load rectifier load output (2 level inverter): Y axis: Rectified output dc voltage, rectified output dc current, Current through diode, voltage across diode; X axis: time (Case II)

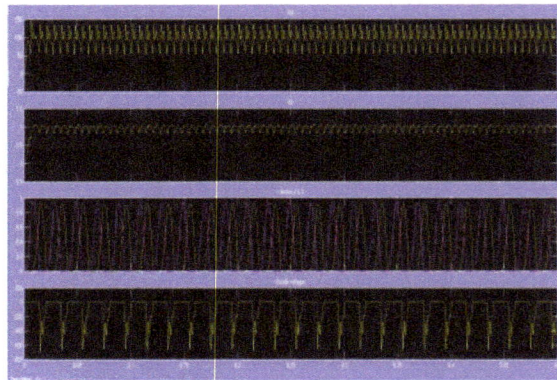

Figure 19 Three-phase load rectifier load output (3 level inverter): Y axis: Rectified output dc voltage, rectified output dc current, Current through diode, voltage across diode; Xaxis: time. (Case III)

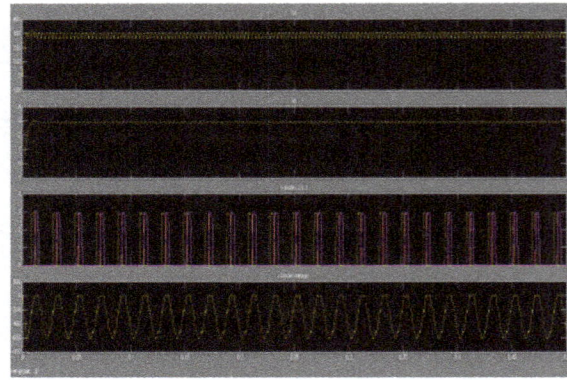

Figure 20 Six-phase load rectifier load output (3 level inverter): Y axis: Rectified output dc voltage, rectified output dc current, Current through diode, voltage across diode; X axis: time. (Case IV)

7. CONCLUSION

This paper presents a quantitative study on reduction of harmonic analysis for six phases as compared to three phases by simulink (MATLAB) for two levels and three levels inverter for non linear loads. Decline in harmonics in load is observed for six phase as compared to three phase load. In addition to harmonic study authors found that ripples of rectified output in case of six phase load circuits is less as compared to its three phase counterpart. As the ratings of various power electronic switches are limited, multilevel voltage source topologies are useful for high voltage and high power applications along with low down harmonics. With the augment of levels of inverter, drop in current THD is observed which further reduces with the increase of three phases to six phases. This implies that there is less requirement of harmonic compensation in case of multiphase load circuit. So, multilevel inverter with multiphase technology pays a promising tool for cost-effective system with effectively reduced harmonics.

REFERENCES

[1] Shantanu Chatterjee. A Multilevel Inverter Based on SVPWM Technique for Photovoltaic Application, *International Journal of Power Electronics and Drive System (IJPEDS)*. 2013; 3(1): 62~73.

[2] Risnidar C, I. Daut, Syafruddin H, N. Hasim, Influence of Harmonics in Laboratory due to Nonlinear Loads, *International Journal of Power Electronics and Drive System (IJPEDS)*. 2012; 2(2): 219~224.

[3] L Srinivas Goud, T Srivani. A Simulation of Three Phase to Multi Phase Transformation using a Special Transformer. *International Journal of Science and Research (IJSR)*, India Online ISSN: 2319-7064. 2013; 2(7).

[4]   Srinivasa Rao Maturu, Avinash Vujji. SVPWM Based Speed Control of Induction Motor Drive with Using V/F Control Based 3-Level Inverter in VSRD *International Journal of Electrical, Electronics and Communication Engineering*. VSRD-IJEECE. 2012; 2(7): 421-437.

[5]   B Somashekar, B Chandrasekhar, David Livingston D. Modeling and Simulation of Three to Nine Phase Using Special Transformer Connection. *International Journal of Emerging Technology and Advanced Engineering*. 2013; 3(6).

[6]   Mika Ikonen, Ossi Laakkonen, Marko Kettunen. Two-level and three-level converter comparison in wind power application. *International Journal of Scientific & Engineering Research*. 2013; 4(1).

[7]   Hag Ahmed Y, Zhengming Zhao, Ting Lu. Comparison and analysis of three-level converters versus two-level converters for ship propulsion applications. *IJRRAS*. 2012; 11(2): 168-175.

[8]   K Ramash Kumar, D Kalyankumar, Dr V Kirubakaran. An Hybrid Multi Level Inverter Based DSTATCOM Control, *Majlesi Journal of Electrical Engineering*. 2011; 5(2).

[9]   Zakir Husain, Ravindra Kumar Singh, Shri Niwas Tiwari. Multi-phase (6-Phase & 12-Phase) Transmission Lines: Performance Characteristics. *International journal of mathematics and computers in simulation*. 2007; 2(1).

[10] GK Singh. Multi-phase induction machine drive research—a survey, Electric Power Systems Research. 2002; 61: 139–147.

[11] Bimal K Bose. Modern Power Electronics and AC Drives, *Prentice Hall of India*, New Delhi. 218-241.

[12] Ovdiu NEAMTU. Three-Phase Inverter with Two Bridges optimized on Simulation. *ACTA Electrotechnica*. 2005; 46(4).

[13] Brendan Peter McGrath, Donald Grahame Holmes. Multicarrier PWM Strategies for Multilevel Inverters. *IEEE transactions on industrial electronics*. 2002; 49(4).

[14] Wenyuan Xu, Jose R Marti, Hermann W Dommel. A multiphase harmonic load flow solution technique. *IEEE transactions on Power Systems*. 1991; 6(1)

[15] Lucian Asiminoaei, Steefan Hansen, Frede Blaabjerg. Predicting Harmonics by Simulations. A Case study for High power Adjustable speed drives. *Electric Power Quality and Utilization*. 2006; 2(1).

[16] Mukhtiar ahmed Mahar, Muhammad Aslamuqaili, And Abdul Sattar Larik. Harmonic Analysis of AC-DC Topologies and their Impacts on Power Systems. *Mehran University Research Journal of Engineering & Technology*. 2011; 30(1) [ISSN 0254-7821]

[17] C Venkatesh, D Srikanth Kumar, DVSS Siva Sarma, M Sydulu, Modelling of Nonlinear Loads and Estimation of Harmonics in Industrial Distribution System. *Fifteenth National Power Systems Conference* (NPSC), IIT Bombay. 2008: 592-597.

[18] Harmonic Current Reduction Techniques for computer systems, *Liebert web notice and conditions* Copyright. Liebert Corporation. 1995-2000.

17

Three-Level DTC Based on Fuzzy Logic and Neural Network of Sensorless DSSM Using Extende Kalman Filter

Elakhdar Benyoussef*, **Abdelkader Meroufel***, **Said Barkat****

*Faculty of Science and Engineering, Department of Electrical Engineering, University of Djilali Liabes, Sidi Bel Abbes 22000, BP 89 Algeria, Intelligent Control Electronic Power System laboratory (I.C.E.P.S)
**Faculty of Technology, Department of Electrical Engineering, University of M'sila, Ichbilia Street, M'sila 28000, BP 166 Algeria

---

ABSTRACT

Keyword:

Direct Torque Control
Double Star
Extended Kalman Filter
Fuzzy Logic Control
Multilevel Inverter
Neural Network
Synchronous Machine

This paper presents a direct torque control is applied for salient-pole double star synchronous machine without mechanical speed and stator flux linkage sensors. The estimation is performed using the extended Kalman filter known by it is ability to process noisy discrete measurements. Two control approaches using fuzzy logic DTC, and neural network DTC are proposed and compared. The validity of the proposed controls scheme is verified by simulation tests of a double star synchronous machine. The stator flux, torque, and speed are determined and compared in the above techniques. Simulation results presented in this paper highlight the improvements produced by the proposed control method based on the extended Kalman filter under various operation conditions.

Corresponding Author:

Elakhdar Benyoussef,
Department of Electrical Engineering,
Djilali Liabes University,
Sidi Bel Abbes 22000, BP 89 Algeria.
Email: lakhdarbenyoussef@yahoo.com

---

1. INTRODUCTION

A multiphase drive has more than three phases in the stator and the same number of inverter legs is in the inverter side. The main advantages of multiphase drives over conventional three-phase drives include increasing the inverter output power, reducing the amplitude of torque ripple and lowering the DC link current harmonics. Furthermore, the multiphase drive system is able to improve the reliability. Indeed, the motor can start and run since the loss of one or many phases [1]. Last two decades, the multiphase drive systems have been used in many applications, such as traction, electric/hybrid vehicles, and ship propulsion [2].

In multiphase machine drive systems, more than threephase windings are implemented in the same stator of the electric machine. One common example of such structure is the double star synchronous motor (DSSM). This motor has two sets of three-phase windings spatially phase shifted by 30 electrical degrees and each set of three-phase stator windings is fed by a three-phase voltage source inverter [3].

The feeding of the DSSM is generally assured by two twolevel inverters. However, for the high power; multilevel inverters are often required. Since the advantages of multilevel inverters and multiphase machines complement each other, it appears to be logical to try to combine them by realizing a multilevel multiphase drive [4]. In the other hand, multilevel inverter fed electric machine systems are considered as a promising approach in achieving high power/high voltage ratings. Moreover, multilevel inverters have the advantages of overcoming voltage limit capability of semiconductor switches, and improving 2 harmonic profiles of output waveforms [5]. The output voltage waveform approaches a sine wave, thus having practically no common-mode voltage and no voltage surge to the motor windings. Furthermore, the reduction in *dv/dt* can prevent motor windings and bearings from failure.

In the other hand, the multilevel direct torque control (DTC) of electrical drives has become an attracting topic in research and academic community over the past decade. Like an every control method has some advantages and disadvantages, DTC method has too. Some of the advantages are presented in [6]. The basic disadvantages of DTC scheme using hysteresis controllers are the variable switching frequency, the current and torque ripple. In the aim to improve the performance of the electrical drives based on traditional DTC, fuzzy logic direct torque control (FLDTC) and artificial neural network direct torque control (DTC-ANN) attracts more and more the attention of many scientists [7], [8]. This paper is devoted to FLDTC and DTC-ANN of sensorless DSSM using extended Kalman filter fed by two three-level diode clamped inverter (DCI).

In this context, several speed observers have been suggested in literature, such as sliding mode observer [9], adaptive observer, model reference adaptive system, and Extende Kalman filter [10]. Kalman filter is a stochastic state observer where nonlinear equations are linearized in every sampling period. It has the advantage of providing both flux and speed estimates, and thus avoids limitations of the open loop pure integration method.

The present paper structure is as follows. Firstly, the model of the DSSM is presented in the second section. In the third section, the three-level inverter modeling is described. In the fourth section, the FLDTC strategy is applied to get decoupled control of the stator flux and electromagnetic torque. Next, a brief introduction on the EKF algorithm is presented in the fifth section. The sixth section introduces the DTC-ANN approach. The seventh section is devoted to the comparative study between three-level FLDTC and three-level DTC-ANN of sensorless DSSM. Finally, conclusions are drawn in the last section.

2. MODELING OF THE DOUBLE STAR SYNCHRONOUS MACHINE

The stator voltages equations are given by:

$$\begin{cases} v_{s1} = R_s i_{s1} + \dfrac{d\phi_{s1}}{dt} \\ v_{s2} = R_s i_{s2} + \dfrac{d\phi_{s2}}{dt} \end{cases} \tag{1}$$

With

v_{s1}, v_{s2} : Stator voltages.

i_{s1}, i_{s2} : Stator currents.

ϕ_{s1}, ϕ_{s2} : Stator flux.

The rotor voltage equation is given by:

$$v_f = R_f i_f + \frac{d\phi_f}{dt} \tag{2}$$

With:

ϕ_f : Flux of rotor excitation.

v_f, i_f : Voltage and current of rotor excitation.

The transformation of the system six phases to the system (α, β) is given by:

$$\begin{bmatrix} X_\alpha & X_\beta \end{bmatrix} = \begin{bmatrix} A \end{bmatrix} \begin{bmatrix} X_{s1} & X_{s2} \end{bmatrix} \tag{3}$$

Where:

X_{s1} and X_{s2} can represent the stator currents, stator flux, and stator voltages.

The transformation matrix A is given by:

$$[A] = \frac{1}{\sqrt{3}} \begin{pmatrix} 1 & -\dfrac{1}{2} & -\dfrac{1}{2} & \dfrac{\sqrt{3}}{2} & -\dfrac{\sqrt{3}}{2} & 0 \\ 0 & \dfrac{\sqrt{3}}{2} & -\dfrac{\sqrt{3}}{2} & \dfrac{1}{2} & \dfrac{1}{2} & -1 \\ 1 & -\dfrac{1}{2} & -\dfrac{1}{2} & -\dfrac{\sqrt{3}}{2} & \dfrac{\sqrt{3}}{2} & 0 \\ 0 & -\dfrac{\sqrt{3}}{2} & \dfrac{\sqrt{3}}{2} & \dfrac{1}{2} & \dfrac{1}{2} & -1 \\ 1 & 1 & 1 & 0 & 0 & 0 \\ 0 & 0 & 0 & 1 & 1 & 1 \end{pmatrix} \tag{4}$$

To express the stator equations in the same reference frame, the following rotation transformation is adopted.

$$P(\theta) = \begin{pmatrix} \cos(\theta) & \sin(\theta) \\ -\sin(\theta) & \cos(\theta) \end{pmatrix} \tag{5}$$

With: θ is the rotor position.

With this transformation, the components of the α-β plane can be expressed in the d-q plane as:

The electrical equations

$$\begin{cases} v_d = R_s i_d + \dfrac{d\phi_d}{dt} - \omega\phi_q \\ v_q = R_s i_q + \dfrac{d\phi_q}{dt} + \omega\phi_d \end{cases} \tag{6}$$

Where:

v_d, v_q : Stator voltages dq components.

i_d, i_q : Stator currents dq components.

ϕ_d, ϕ_q : Stator flux dq components.

The flux equations

$$\begin{cases} \phi_d = L_d i_d + M_{fd} i_f \\ \phi_q = L_q i_q \\ \phi_f = L_f i_f + M_{fd} i_d \end{cases} \tag{7}$$

The mechanical equation

$$J\frac{d\Omega}{dt} = T_{em} - T_L - f\,\Omega \tag{8}$$

With:

T_{em}, T_L : Electromagnetic and load torque.

Ω : Rotor speed.

The electromagnetic torque

$$T_{em} = p\left(\phi_d i_q - \phi_q i_d\right) \tag{9}$$

3.   MODELING OF THE THREE-LEVEL INVERTER

Figure 1 shows the circuit of a three-level diode clamped inverter and the switching states of each leg of the inverter. Each leg is composed of two upper and lower switches with anti-parallel diodes. Two series DC-link capacitors split the DC-bus voltage in half, and six clamping diodes confine the voltage across the switches within the voltage of the capacitors, each leg of the inverter can have three possible switching states; 2, 1 or 0 [11].

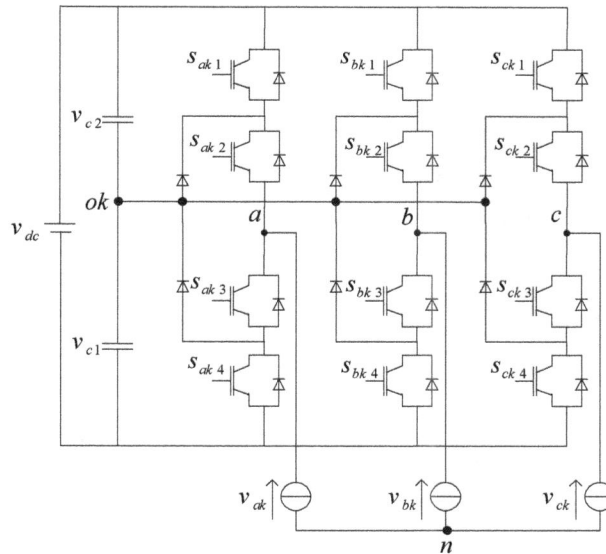

Figure 1. Schematic diagram of a three-level inverter (k=1 for first inverter and k=2 for second inverter)

The representation of the space voltage vectors of a three-level inverter for all switching states is given by Figure 2. According to the magnitude of the voltage vectors, the voltage vectors can be partitioned into four groups: the zero voltage vectors v_0, the large voltage vectors (v_{1L}, v_{3L}, v_{5L}, v_{7L}, v_{9L}, v_{11L}), the middle voltage vectors (v_{2L}, v_{4L}, v_{6L}, v_{8L}, v_{10L}, v_{12L}), and the small voltage vectors (v_{1S}, v_{2S}, v_{3S}, v_{4S}, v_{5S}, v_{6S}).

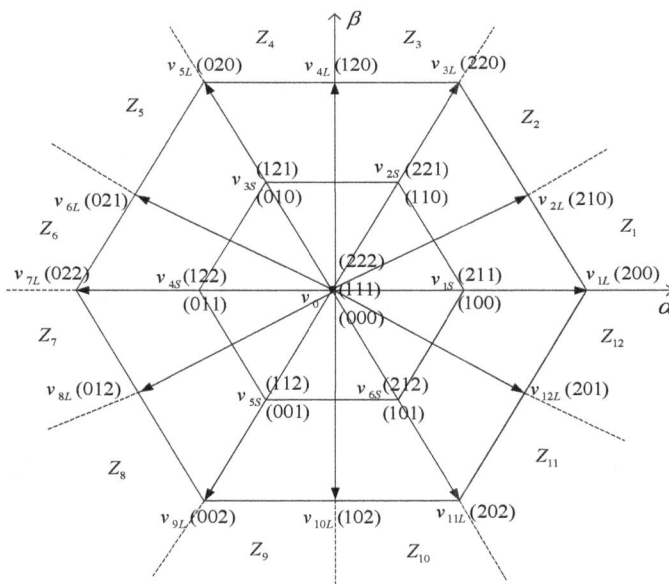

Figure 2. Space vector diagram of three-level inverter

4.   DIRECT TORQUE CONTROL BASED ON FUZZY LOGIC STRATEGY

The principle of fuzzy logic direct torque control is similar to traditional DTC. However, the hysteresis controllers are replaced by fuzzy controller and the output vector of the fuzzy controller is led to a switching table to decide which vector should be applied. This method based on fuzzy classification has the advantage of simplicity and easy implementation [8].

The components of stator flux can be estimated by:

$$
\begin{cases}
\hat{\phi}_\alpha(t) = \int_0^t (\hat{v}_\alpha - R_s i_\alpha) d\tau + \hat{\phi}_\alpha(0) \\
\hat{\phi}_\beta(t) = \int_0^t (\hat{v}_\beta - R_s i_\beta) d\tau + \hat{\phi}_\beta(0)
\end{cases}
\tag{10}
$$

The stator flux amplitude is given by:

$$
\left|\hat{\phi}_s\right| = \sqrt{\hat{\phi}_\alpha^2 + \hat{\phi}_\beta^2}
\tag{11}
$$

The stator flux angle is calculated by:

$$
\hat{\theta}_s = \tan 2^{-1}\left(\frac{\hat{\phi}_\beta}{\hat{\phi}_\alpha}\right)
\tag{12}
$$

Electromagnetic torque equation is given by:

$$
\hat{T}_{em} = p\left(\hat{\phi}_\alpha i_\beta - \hat{\phi}_\beta i_\alpha\right)
\tag{13}
$$

The fuzzy controller is designed to have three fuzzy state variables and one control variable for achieving constant torque and flux control. The first variable $(E_\phi = \hat{\phi}_s - \phi_s^*)$ is the difference between the command stator flux ϕ_s^* and the estimated stator flux magnitude $\hat{\phi}_s$. The second variable $(E_T = \hat{T}_{em} - T_{em}^*)$ is the difference between the command electric torque T_{em}^* and estimated electric torque $\hat{T}_{em}$. The third fuzzy state variable is the stator flux phase $(\hat{\theta}_s)$. Figure 3 gives the membership functions for input variables E_ϕ, E_T and $\hat{\theta}_s$.

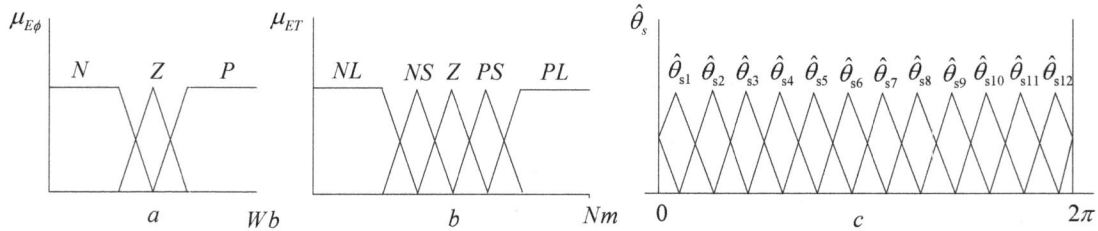

Figure 3. Membership functions of input variables: a) Stator flux error, b) Torque error, c) Stator flux angle

The switching tables of the proposed three-level FLDTC are used to select the best output voltage depending on the position of the stator flux and desired action on the torque and stator flux. The optimal voltage vector selection, for controlling both the amplitude and rotating direction of the stator flux, is indicated in Table 1, 2. The linguistic terms used for stator flux error are N (negative), Z (zero), and P (positive). For the torque error, the terms used are NL (negative large), NS (negative small), ZE (zero), PS (positive small), and PL (positive large).

Table 1. Rules of fuzzy control for first star

$\hat{\theta}_{s1}$				$\hat{\theta}_{s2}$				$\hat{\theta}_{s3}$			
E_T　E_ϕ	P	Z	N	E_T　E_ϕ	P	Z	N	E_T　E_ϕ	P	Z	N
PL	v_{3L}	v_{2S}	v_{5L}	PL	v_{3L}	v_{2S}	v_{5L}	PL	v_{5L}	v_{3S}	v_{7L}
PS	v_{2L}	v_{2S}	v_{4L}	PS	v_{4L}	v_{3S}	v_{6L}	PS	v_{4L}	v_{3S}	v_{6L}
ZE	0	0	0	ZE	0	0	0	ZE	0	0	0
NS	v_{12L}	0	v_{10L}	NS	v_{12L}	0	v_{10L}	NS	v_{2L}	0	v_{12L}
NL	v_{11L}	v_{5S}	v_{9L}	NL	v_{1L}	v_{6S}	v_{11L}	NL	v_{1L}	v_{6S}	v_{11L}

$\hat{\theta}_{s4}$				$\hat{\theta}_{s5}$				$\hat{\theta}_{s6}$			
E_T　E_ϕ	P	Z	N	E_T　E_ϕ	P	Z	N	E_T　E_ϕ	P	Z	N
PL	v_{5L}	v_{3S}	v_{7L}	PL	v_{7L}	v_{4S}	v_{9L}	PL	v_{7L}	v_{4S}	v_{9L}
PS	v_{6L}	v_{4S}	v_{8L}	PS	v_{6L}	v_{4S}	v_{8L}	PS	v_{8L}	v_{5S}	v_{10L}
ZE	0	0	0	ZE	0	0	0	ZE	0	0	0
NS	v_{2L}	0	v_{12L}	NS	v_{4L}	0	v_{2L}	NS	v_{4L}	0	v_{2L}
NL	v_{10L}	v_{1S}	v_{1L}	NL	v_{3L}	v_{1S}	v_{1L}	NL	v_{5L}	v_{2S}	v_{3L}

$\hat{\theta}_{s7}$				$\hat{\theta}_{s8}$				$\hat{\theta}_{s9}$			
E_T　E_ϕ	P	Z	N	E_T　E_ϕ	P	Z	N	E_T　E_ϕ	P	Z	N
PL	v_{9L}	v_{5S}	v_{11L}	PL	v_{9L}	v_{5S}	v_{11L}	PL	v_{11L}	v_{6S}	v_{1L}
PS	v_{8L}	v_{5S}	v_{10L}	PS	v_{10L}	v_{6S}	v_{12L}	PS	v_{10L}	v_{6S}	v_{12L}
ZE	0	0	0	ZE	0	0	0	ZE	0	0	0
NS	v_{6L}	0	v_{4L}	NS	v_{6L}	0	v_{4L}	NS	v_{8L}	0	v_{6L}
NL	v_{5L}	v_{2S}	v_{3L}	NL	v_{7L}	v_{3S}	v_{5L}	NL	v_{7L}	v_{3S}	v_{5L}

$\hat{\theta}_{s10}$				$\hat{\theta}_{s11}$				$\hat{\theta}_{s12}$			
E_T　E_ϕ	P	Z	N	E_T　E_ϕ	P	Z	N	E_T　E_ϕ	P	Z	N
PL	v_{11L}	v_{6S}	v_{1L}	PL	v_{1L}	v_{1S}	v_{3L}	PL	v_{1L}	v_{1S}	v_{3L}
PS	v_{12L}	v_{1S}	v_{2L}	PS	v_{12L}	v_{1S}	v_{2L}	PS	v_{2L}	v_{2S}	v_{4L}
ZE	0	0	0	ZE	0	0	0	ZE	0	0	0
NS	v_{8L}	0	v_{6L}	NS	v_{10L}	0	v_{8L}	NS	v_{10L}	0	v_{8L}
NL	v_{9L}	v_{4S}	v_{7L}	NL	v_{9L}	v_{4S}	v_{7L}	NL	v_{11L}	v_{5S}	v_{9L}

Table 2. Rules of fuzzy control for second star

$star1$	$\hat{\theta}_{s12}$	$\hat{\theta}_{s1}$	$\hat{\theta}_{s2}$	$\hat{\theta}_{s3}$	$\hat{\theta}_{s4}$	$\hat{\theta}_{s5}$	$\hat{\theta}_{s6}$	$\hat{\theta}_{s7}$	$\hat{\theta}_{s8}$	$\hat{\theta}_{s9}$	$\hat{\theta}_{s10}$	$\hat{\theta}_{s11}$
$star2$	$\hat{\theta}_{s1}$	$\hat{\theta}_{s2}$	$\hat{\theta}_{s3}$	$\hat{\theta}_{s4}$	$\hat{\theta}_{s5}$	$\hat{\theta}_{s6}$	$\hat{\theta}_{s7}$	$\hat{\theta}_{s8}$	$\hat{\theta}_{s9}$	$\hat{\theta}_{s10}$	$\hat{\theta}_{s11}$	$\hat{\theta}_{s12}$

5.　SPEED ESTIMATION BASED ON EXTENDED KALMAN FILTER

Kalman filter is a state observer that establishes the best approximation by minimization of the square error for the state variables of a system, subjected at both its input and output to random disturbances. If the dynamic system of which the state is being observed is nonlinear, then the Kalman filter is called an extended one (EKF) [10]. The development of the Kalman filter is closely linked to the stochastic systems. The linear stochastic systems are described by:

$$\begin{cases} \dot{x}(t) = Ax(t) + Bu(t) + w(t),\ x(t_o) = x_o \\ y(t) = Cx(t) + v(t) \end{cases} \tag{14}$$

Where: w and v are the system and measurement noise.

Estimation of an error covariance matrix:

$$P^-(k+1) = A_d P(k) A_d^T + Q \tag{15}$$

Computations of a Kalman filter gain:

$$K(k+1) = P^-(k+1)C^T \left(CP^-(k+1)C^T + R\right)^{-1} \tag{16}$$

Update of an error covariance matrix:

$$P(k+1) = \left(I - K(k+1)C\right)P^-(k+1) \tag{17}$$

State estimation:

$$\hat{x}(k+1) = \hat{x}(k) + K(k+1)\left(y(k+1) - C.\hat{x}(k+1)\right) \tag{18}$$

Where:

$P^-(k+1)$: is a priori error covariance matrix Q and R respectively.

The extended Kalman filter implementation for a DSSM requires three basic steps:
a)   Continuous DSSM model
b)   Discretization of the DSSM model
c)   Simulation

5.1. Continuous DSSM Model
The model of DSSM in the α-β reference can be written in the following from:

$$\begin{cases} \dot{x}(t) = Ax(t) + Bu(t) \\ y(t) = Cx(t) \end{cases} \tag{19}$$

With:

$$x(t) = \begin{bmatrix} i_\alpha & i_\beta & \phi_\alpha & \phi_\beta & \Omega & \theta \end{bmatrix}$$
$$y = \begin{bmatrix} i_\alpha & i_\beta \end{bmatrix}, \; u = \begin{bmatrix} v_\alpha & v_\beta \end{bmatrix}$$

$$A = \begin{bmatrix} -R_s & -p\Omega L_q & 0 & p\Omega & 0 & 0 \\ p\Omega L_q & -R_s & -p\Omega & 0 & 0 & 0 \\ -R_s & 0 & 0 & 0 & 0 & 0 \\ 0 & -R_s & 0 & 0 & 0 & 0 \\ -p\phi_\beta/J & p\phi_\alpha/J & 0 & 0 & -f/J & 0 \\ 0 & 0 & 0 & 0 & p & 0 \end{bmatrix}, \; B = \begin{bmatrix} 1/L_q & 0 & 0 \\ 0 & 1/L_q & 0 \\ 1 & 0 & 0 \\ 0 & 1 & 0 \\ 0 & 0 & -f/J \\ 0 & 0 & 0 \end{bmatrix}$$

With:

$$\begin{cases} \phi_\alpha = L_q i_\alpha + \phi_f \cos(\theta) \\ \phi_\beta = L_q i_\beta + \phi_f \sin(\theta) \end{cases}$$

5.2. Discretization of the DSSM Model
The corresponding discrete time model is given by:

$$\begin{cases} x_{(k+1)} = A_d x_{(k)} + B_d u_{(k)} \\ y_{(k+1)} = C_d x_{(k)} \end{cases} \qquad (20)$$

The conversion is done by the following approximation:

$$\begin{cases} A_d = e^{At} = I + AT_s \\ B_d = \int_0^t e^{A\xi} B d\xi = BT_s \\ C_d = C \end{cases} \qquad (21)$$

5.3. Simulation

Based on the previous elements, the EKF can now be built and applied to the DSSM. By derivation of the vector function in relation to the state vector, matrix A_d, B_d and C_d are given by:

$$A_d = \begin{bmatrix} 1-T_sR_s/L_q & -T_sp\Omega L_q & 0 & T_sp\Omega & 0 & 0 \\ T_sp\Omega L_q & 1-T_sR_s/L_q & -T_sp\Omega & 0 & 0 & 0 \\ -T_sR_s & 0 & 0 & 0 & 0 & 0 \\ 0 & -T_sR_s & 1 & 0 & 0 & 0 \\ -T_sp\phi_\beta/J & T_sp\phi_\alpha/J & 0 & 1 & -T_sf/J & 0 \\ 0 & 0 & 0 & 0 & T_sp & 1 \end{bmatrix}, B_d = \begin{bmatrix} T_s/L_q & 0 & 0 \\ 0 & T_s/L_q & 0 \\ T_s & 0 & 0 \\ 0 & T_s & 0 \\ 0 & 0 & -T_sf/J \\ 0 & 0 & 0 \end{bmatrix}, C_d = \begin{bmatrix} 1 & 0 & 0 & 0 & 0 \\ 0 & 1 & 0 & 0 & 0 \end{bmatrix}$$

The structure of DTC based on fuzzy logic control of DSSM is shown in Figure 4.

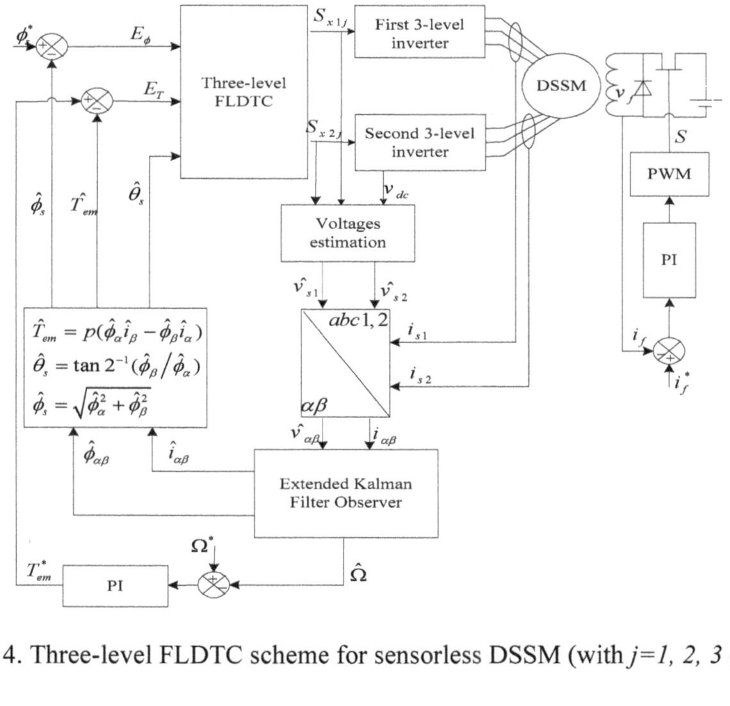

Figure 4. Three-level FLDTC scheme for sensorless DSSM (with *j=1, 2, 3* or *4*)

6.   DIRECT TORQUE CONTROL BASED ON NEURAL NETWORK STRATEGY

The ANN has many models; but the usual model is the multilayer feed forward net work using the error back propagation algorithm [9]. Such a neural network contains three layers: input layers, hidden layers and output layers (Figure 5). Each layer is composed of several neurons. The number of the neurons in the input and output layers depends on the number of the selected input and output variables. The number of hidden layers and the number of neurons in each depend on the desired degree of accuracy.

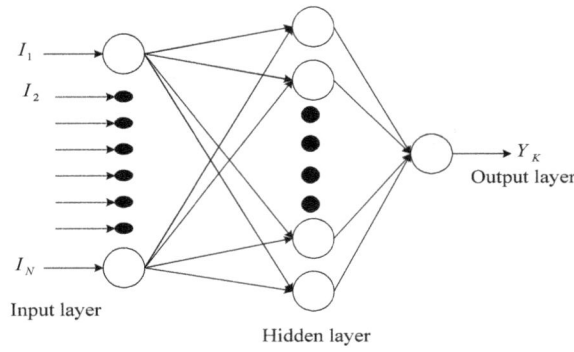

Figure 5. Architecture of Multilayer Neural Network

The structure of the neural network to perform the DTC applied to DSSM satisfactorily was a neural network with 3 linear input nodes, 12 neurons in the hidden layer, and 6 neurons in the output layer, as shown in Figure 6.

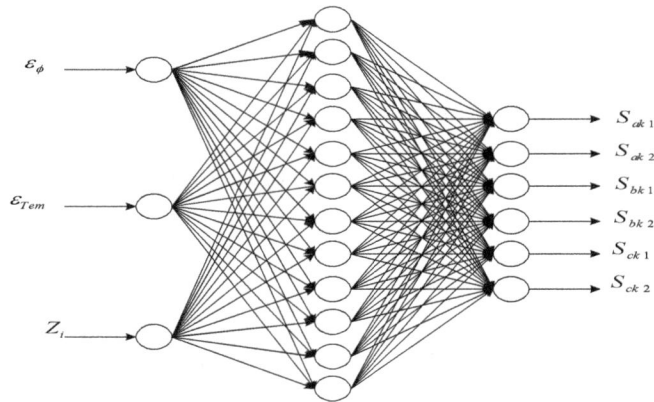

Figure 6. Neural network structure for three-level DTC

The general structure of the DSSM with DTC-ANN using a three-level inverter in each star is represented by Figure 7.

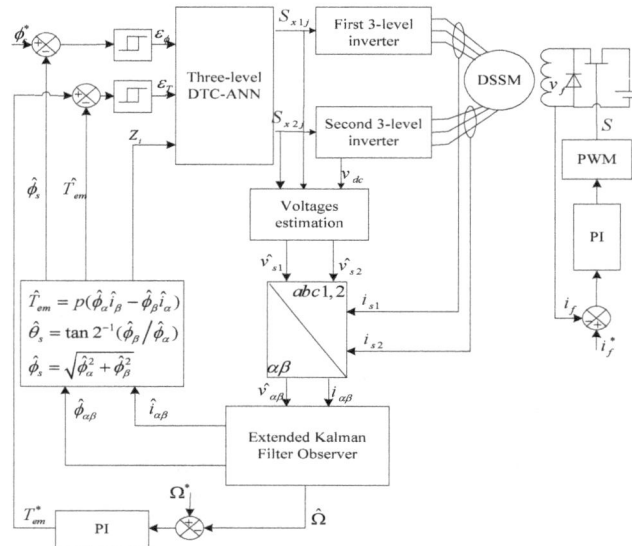

Figure 7. Three-level DTC-ANN scheme for sensorless DSSM (with $j=1, 2, 3$ or 4)

7. SIMULATION RESULTS

The proposed controls based on EKF observer are tested by some numerical simulations to verify its effectiveness in the steady-state and dynamic. Parameters of EKF are chosen as following to ensure filter not divergent.

$$[Q] = \begin{bmatrix} 100 & 0 & 0 & 0 & 0 & 0 \\ 0 & 100 & 0 & 0 & 0 & 0 \\ 0_s & 0 & 0.1 & 0 & 0 & 0 \\ 0 & 0 & 0 & 0.1 & 0 & 0 \\ 0 & 0 & 0 & 0 & 0.1 & 0 \\ 0 & 0 & 0 & 0 & 0 & 0.1 \end{bmatrix}, [P] = \begin{bmatrix} 0.1 & 0 & 0 & 0 & 0 & 0 \\ 0 & 0.1 & 0 & 0 & 0 & 0 \\ 0 & 0 & 10^{-5} & 0 & 0 & 0 \\ 0 & 0 & 0 & 10^{-5} & 0 & 0 \\ 0 & 0 & 0 & 0 & 10^{-5} & 0 \\ 0 & 0 & 0 & 0 & 0 & 0.1 \end{bmatrix}, [R] = \begin{bmatrix} 0.12 & 0 \\ 0 & 0.12 \end{bmatrix}$$

The simulation results of three-level DTC-ANN of sensorless DSSM are compared with three-level FLDTC. For this end, the controls system was tested under deferent operating conditions such as sudden change of load torque and step change in reference speed.

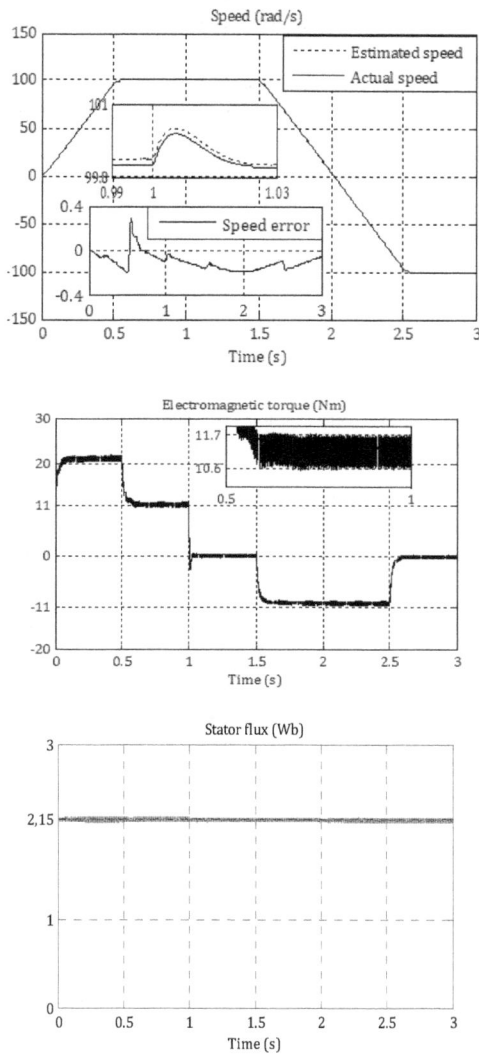

Figure 8. Dynamic responses of three-level FLDTC for sensorless DSSM

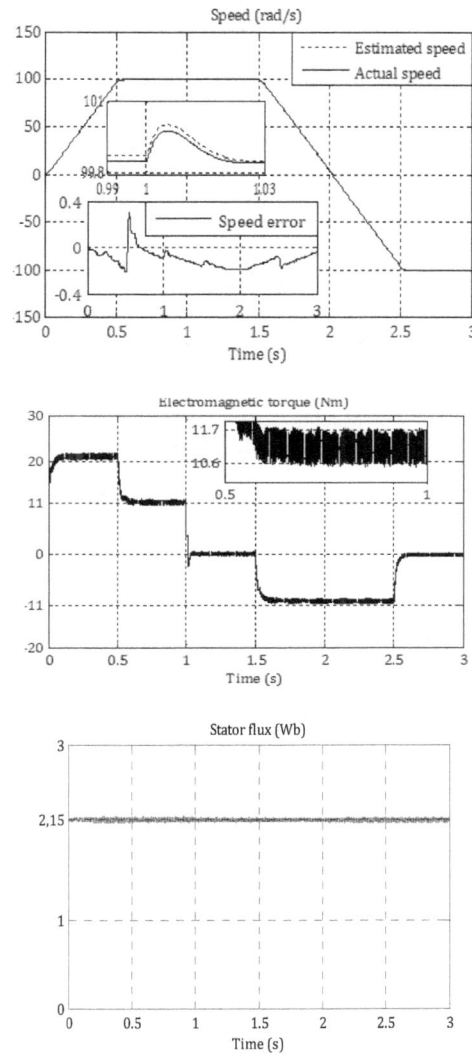

Figure 9. Dynamic responses of three-level DTC-ANN for sensorless DSSM

The obtained results are presented in Figure 8 for the three-level FLDTC and Figure 9 for the three-level DTC-ANN of DSSM without speed sensor. The DSSM is accelerating from standstill to reference speed *100* rad/s. The system is started with full load torque (T_L=*11* N.m). Afterwards, a step variation on the load torque (T_L=*0* N.m) is applied at time *t*=*1*s. And then a sudden reversion in the speed command from *100* rad/s to *-100* rad/s was introduced at *1.5* s.

Figure 8 and 9 depict that the speed response is merged with the reference one and the flux is constant equal to its rated value. As it can be seen, the employment of three-level DTC-ANN without speed sensor permits to obtain the same dynamic performances as those obtained with a three-level FLDTC without speed sensor. Indeed the observed and actual speed responses are closed to their reference without any overshoot and steady-state error.

The obtained results shown that the three-level DTC-ANN sensorless DSSM ensures good decoupling between stator flux linkage and electromagnetic torque. Also, it can decrease the torque ripples in comparison to the three-level FLDTC sensorless DSSM.

8.   Conclusion

In this paper, a direct torque control based on artificial neural network and fuzzy logic methods applied of sensorless double star synchronous machine using extended Kalman filter is presented. It is pointed out that the robustness of the controlled double star synchronous machine drive against speed and load torque variations is guaranteed. Furthermore, the DTC-ANN control scheme decreases considerably the electromagnetic torque ripples and assures good speed tracking without overshoot. The decoupling between the stator flux and the electromagnetic torque is maintained, confirming the good dynamic performances of the developed multiphase drive system. Also, the robustness of the observer, against speed and load variations confirms the good dynamic performances of the developed sensorless multiphase drive.

References

[1]   L Nezli, M Mahmoudi. Vector Control with Optimal Torque of a Salient–Pole Double Star Synchronous Machine Supplied by Three–Level Inverters. *Journal of Electrical Engineering*. 2010; 61(5): 257-263.

[2]   A Matyas, A BiroK, D Fodorean. Multi-phase Synchronous Motor Solution for Steering Applications. *Progress in Electromagnetic Research*. 2012; 131: 63-80.

[3]   S Kallio, M Andriollo, A Tortella, J Karttunen. Decoupled d–q Model of Double-Star Interior-Permanent-Magnet Synchronous Machines. *IEEE Transactions on Industrial Electronics*. 2013; 60(6): 2486-2494.

[4]   E Benyoussef, A Meroufel, S Barkat. Multilevel Direct Torque Balancing Control of Double Star Synchronous Machine. *Journal of Electrical Engineering*. 2014; 1-11.

[5]   B Singh, N Mittal, D Verma, D Singh, S Singh, R Dixit, M Singh, A Baranwal. Multi-Level Inverter: a Literature Survey on Topologies and Control Strategies. *International Journal of Reviews in Computing*. 2012; 10: 1-16.

[6]   B Naas, L Nezli, M Mahmoudi, M Elbar. Direct Torque Control Based Three-Level Inverter Fed Double Star Permanent Magnet Synchronous Machine. *Energy Procedia*. 2012; 18: 521-530.

[7]   F Kadri, S Drid, D Djarah, F Djeffal. Direct Torque Control of Induction Motor Fed by Three Phase PWM Inverter Using Fuzzy Logic and Neural Network. *Sixth International Conference on Electrical Engineering*. 2010; 12-16.

[8]   A Tessarolo, D Giulivo. Direct Torque Control of Induction Motors with Fuzzy Logic Controller. *International Symposium on Power Electronics, Electrical Drives, Automation and Motion*. 2010; 845-852.

[9]   X Zhuang, M Rahman. Comparison of a Sliding Observer and a Kalman Filter for Direct Torque Controlled IPM Synchronous Motor Drives. *IEEE, Transactions on Industrial Electronics*. 2012; 59(11): 4179-4188.

[10] M Merzoug, H Benalla, H Naceri. Speed Estimation Using Extended Filter Kalman for the Direct Torque Controlled Permanent Magnet Synchronous Motor (PMSM). *IEEE, International Conference on Computer and Electrical Engineering*, Dubai. 2009; 124-127.

[11] H Kalpesh, P Agarwal. Space Vector Modulation with DC Link Voltage Balancing Control for Three Level Inverters. *International Journal of Recent Trends in Engineering*. 2009; 1(3): 229-233.

18

Effect of Parametric Variations and Voltage Unbalance on Adaptive Speed Estimation Schemes for Speed Sensorless Induction Motor Drives

Mohan Krishna. S, Febin Daya. J.L
School of Electrical Engineering, VIT University, Chennai, India

---

ABSTRACT

Keyword: Adaptive control Adaptive speed observers Machine model Model reference Speed estimation	Speed Estimation without speed sensors is a complex phenomenon and is overly dependent on the machine parameters. It is all the more significant during low speed or near zero speed operation. There are several approaches to speed estimation of an induction motor. Eventually, they can be classified into two types, namely, estimation based on the machine model and estimation based on magnetic saliency and air gap space harmonics. This paper analyses the effect of incorrect setting of parameters like the stator resistance, rotor time constant, load torque variations and also Voltage unbalance on various adaptive control based speed estimation techniques fed from the machine model. It also shows how the convergence mechanisms of the adaptation schemes are affected during these conditions. The equivalent models are built and simulated offline using MATLAB/SIMULINK blocksets and the results are analysed.

Corresponding Author:

Mohan Krishna.S,
School of Electrical Engineering (SELECT),
VIT University, Chennai Campus, Chennai - 600 127, India.
Email: smk87.genx@gmail.com

1. INTRODUCTION

The essence of employing encoderless induction motor drives is to eliminate additional space and cost which would otherwise be attributed to the speed encoder. The use of speed encoders also acts contrary to the inherent robustness of the induction motors. Therefore, estimation of speed without speed sensors emerged as an important concept [1]. Great amount of research has been done in this regard and it continues to inspire more, with the onset of artificial intelligence based speed estimation and other emerging technologies. The speed can be estimated either from the magnetic saliencies or by a machine model fed by terminal quantities. Owing to the complexity of speed estimation, the most discussed problems were the estimator's sensitivity to motor parameter changes, low and zero speed operation, speed estimation at field weakening region, stability problems in the regenerative mode etc.

This paper attempts to present a performance analysis of various adaptive control schemes when they are subjected to load perturbations, incorrect parameter settings (Stator resistance and Rotor time constant) and Voltage unbalance. The effect of the same on the convergence of the adaptive mechanism is also presented.

2. MATHEMATICAL MODEL OF THE INDUCTION MOTOR

The dynamic state space model of the induction motor is presented below, which, aids in the formulation of estimation and control algorithms. It also helps in determining the internal behavior of the system along with the desired input and output. The stator current and the rotor flux are the state variables [2]:

$$p \begin{bmatrix} i_s^e \\ \psi_r^e \end{bmatrix} = \begin{bmatrix} A_{11} & A_{12} \\ A_{21} & A_{22} \end{bmatrix} \begin{bmatrix} i_s^e \\ \psi_r^e \end{bmatrix} + \begin{bmatrix} B_{11} \\ 0 \end{bmatrix} V_s^e \tag{1}$$

$$\dot{x}^e = A x^e + B v_s^e \tag{2}$$

$$i_s^e = C x^e \tag{3}$$

Where,

$$x^e = [i_s^e \ \psi_r^e]^T, i_s^e = [i_{ds}^e \ i_{qs}^e]^T, \psi_r^e = [\psi_{dr}^e \ \psi_{qr}^e]^T \tag{4}$$

The electromechanical torque is given by,

$$T_e = \frac{3}{2} \frac{P}{2} \frac{L_m}{L_r} \left(i_{qs}^e \psi_{dr}^e - i_{ds}^e \psi_{qr}^e \right), \tag{5}$$

3.   INDUCTION MOTOR FOC AND SPEED ESTIMATION

In the conventional scalar control of Induction motor, since torque and flux linkages are a function of voltage, current or frequency, there is an inherent coupling present due to which the response is sluggish. Therefore, there is a need to decouple the same, by making the torque independent of flux. This is known as vector control or field oriented control of the Induction motor. This is similar to the orthogonal orientation of the flux and torque achieved in a separately excited dc motor [3]. Generally, the stator current is resolved into the torque producing component and flux producing component. The DC machine like performance is only possible if the flux producing component of the current is oriented in the direction of flux and the Torque component of the current is perpendicular to it. The orientation is possible with either the rotor flux (ψ_r), airgap flux (ψ_m) or stator flux (ψ_s). However, rotor flux oriented control gives natural decoupling effect, whereas airgap or stator flux orientation have coupling in the flux control loop. The Figure 1. Shows the different types of Field Oriented control.

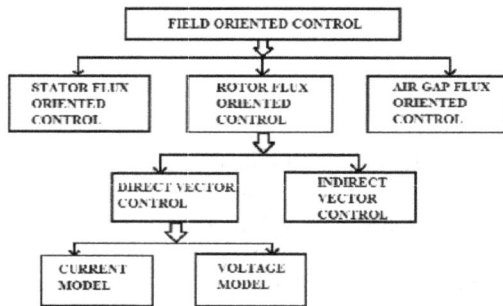

Figure 1. Field Oriented Control schemes for Induction Motor

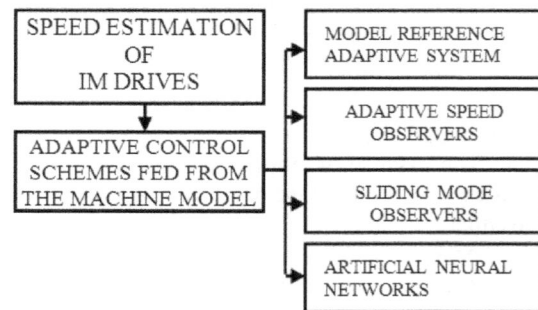

Figure 2. Classification of Speed estimation methods

Figure 2 illustrates the different types of adaptive control schemes fed from the terminal quantities of the machine. These methods display good performance at high and medium speeds. But they are not stable at very low operating speeds as they are parameter dependent and parameter errors can degrade speed performance. The prominent configurations of adaptive speed estimation schemes are presented below.

3.1. Model Reference Adaptive Control (MRAC)

As the name suggests, an adaptive system adapts itself to the controlled system with parameters which need to be estimated or are uncertain. Unlike robust control, it does not need any first hand information about the bounds on these estimated or uncertain parameters. The primary aim of adaptive control is parameter estimation. The MRAS forms the crux of adaptive control. The MRAS is easy to implement and has a high speed of convergence and adaptation and it also displays robust performance to parameter variations. The general configuration of MRAS is shown in Figure 3. The error vector is obtained as the difference in the outputs of the reference and adjustable models. The two models are fed from the machine terminals. The adaptive mechanism forces the error vector to zero in order to converge the estimated output to the reference output [4]. During the design of the adaptive control scheme, special emphasis has to be laid on the convergence mechanism.Since stability of the estimator is of great concern at all speeds, Lyapunov stability criterion plays an important role in deriving the control laws and force convergence as well as ensure fast error dynamics. Adaptive mechanisms can be in the form of fixed gain PI Regulators, Fuzzy Logic (FL), Sliding Mode (SM) based etc. As Sensorless Model based speed estimation methods are sensitive to machine parameters, several methods and algorithms have been proposed for parameter adaptation also [5], in order to optimise the performance of the drive etc. MRAS based approach varies with the quantity that is selected as output of the reference and adjustable model [6]. The more popular choices happen to be rotor flux, back emf, stator currents and Instantaneous reactive power [7].

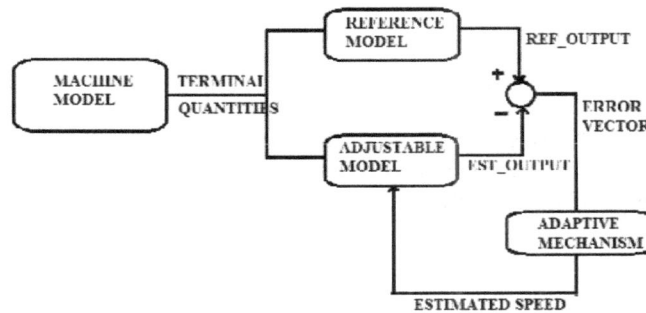

Figure 3. General Configuration of MRAS

The following Equation (6) – (10) are used to characterize the rotor flux based MRAS speed estimator along with the adaptive mechanism used [8]:

a) Reference Model:

$$\psi_{qr}^{s} = L_r/L_m[\int(V_{qs}^{s}-R_si_{qs}^{s}-\sigma L_si_{qs}^{s})dt] \tag{6}$$

$$\psi_{dr}^{s} = L_r/L_m[\int(V_{ds}^{s}-R_si_{ds}^{s}-\sigma L_si_{ds}^{s})dt] \tag{7}$$

Where $\sigma = 1 - L_m^2/L_sL_r$

b) Adjustable Model:

$$d\psi_{qr}^{s}/dt = -1/T_r\,\psi_{qr}^{s} + \omega_r\,\psi_{dr}^{s} + L_m/T_r\,i_{qs}^{s} \tag{8}$$

$$d\psi_{dr}^{s}/dt = -1/T_r\,\psi_{dr}^{s} - \omega_r\,\psi_{qr}^{s} + L_m/T_r\,i_{ds}^{s} \tag{9}$$

c) Adaptive Mechanism:

$$\widehat{\omega}_r = \left\{ K_P + \frac{K_I}{p}\right\}(\varphi_q\,\widehat{\varphi}_d - \varphi_d\,\widehat{\varphi}_q) \tag{10}$$

3.2. Adaptive Speed Observers

H.Kubota et al, [9] proposed a Full order speed Adaptive Flux Observer (AFFO) based on adaptive control theory. The AFFO stabilises the performance of the drive even at low speed region by allocating poles arbitrarily. It makes use of either the Lyapunov's stability criterions or the Popov's criterions to derive the adaptive scheme. The AFFO, apart from estimating the Stator current and rotor flux, also makes use of a gain matrix which is used for stability purpose. The general configuration of the observer is shown in Figure 4.

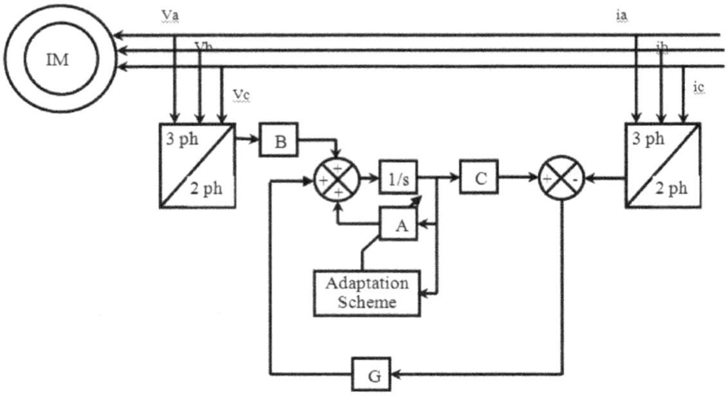

Figure 4. Adaptive Observer scheme for speed estimation

Where, 'A' is the system matrix, the symbol '^' indicates estimated values, 'X' comprises the state variables which are the direct and quadrature axes stator currents and rotor fluxes, 'G' is the observer gain matrix, chosen in such a way that the Eigen values of the observer are proportional to the Eigen values of the machine to ensure stability under normal operating condition. The state equations depicting the structure of the Adaptive Pseudo reduced order speed observer (AFFO) is shown [10]:

(a) Reference Model (Motor model):

$$\frac{dx}{dt} = [A]x + [B]u \tag{11}$$

$$y = [C]x \tag{12}$$

Where,

$$x = [i_{ds}, i_{qs}, \varphi_{dr}, \varphi_{qr}]^T, \quad A = \begin{bmatrix} A_{11} & A_{12} \\ A_{21} & A_{22} \end{bmatrix}, I = \begin{bmatrix} 1 & 0 \\ 0 & 1 \end{bmatrix}, J = \begin{bmatrix} 0 & -1 \\ 1 & 0 \end{bmatrix}$$

$$A_{11} = -\left[\frac{R_s}{\sigma L_s} + \frac{1-\sigma}{\sigma T_r}\right]I = a_{r11}I, A_{12} = \frac{L_m}{\sigma L_s L_r}\left[\frac{1}{T_r}I - \omega_r J\right] = a_{r12}I + a_{i12}J, A_{21} = \frac{L_m}{T_r}I = a_{r21}I,$$

$$A_{22} = \frac{-1}{T_r}I + \omega_r J = a_{r22}I + a_{i22}J$$

$$B = [\frac{1}{\sigma L_s}I \quad 0]^T, \quad C = [I, 0], u = [V_{ds} \ V_{qs}]^T$$

(b) Adjustable Model (Luenberger Adaptive Speed Observer):

$$\frac{d\hat{x}}{dt} = [\widehat{A}]\hat{x} + [B]u + [G](\hat{\imath}_s - i_s) \tag{13}$$

$$\hat{y} = [C]\hat{x} \tag{14}$$

Where, $\hat{\imath}_s$ = estimated value of stator current and,
i_s = measured value of stator current

$$A = \begin{bmatrix} A_{11} & \widehat{A}_{12} \\ A_{21} & \widehat{A}_{22} \end{bmatrix}$$

$$\widehat{A}_{12} = \frac{L_m}{\sigma L_s L_r}\left[\frac{1}{T_r}I - \widehat{\omega}_r J\right] = a_{r12}I + \hat{a}_{i12}J, \quad \widehat{A}_{22} = \frac{-1}{T_r}I + \widehat{\omega}_r J = a_{r22}I + \hat{a}_{i22}J$$

The term $[G](\hat{\imath}_s - i_s)$ is used as a correction term for the Adaptive Speed Observer. 'G' is the reduced order observer gain matrix designed so as to make (13) stable. The pseudo reduced order gain matrix is chosen as follows [11], [12]:

$$G = \begin{bmatrix} g_1 & g_2 \\ -g_2 & g_1 \end{bmatrix}^T \tag{15}$$

The observer gain matrix is calculated based on the pole placement technique, so that the state of the observer will converge to the reference model (the motor). Therefore, the eigen values are chosen relatively more negative than the eigen values of the reference model, so as to ensure faster convergence. It is chosen as follows:

$$g_1 = (k-1)a_{r11}, g_2 = k_p, k_p \geq -1$$

Where, g_1 depends on the motor parameters, g_2 and k_p are arbitrarily chosen and k is an arbitrary positive constant.

(c) Adaptive Mechanism:
The Lyapunov function candidate defined for the adaptation scheme is [10]:

$$V = e^T e + \frac{(\hat{\omega}_r - \omega_r)^2}{\lambda} \tag{16}$$

Where λ is a positive constant.
The time derivative of 'V' becomes,

$$\frac{dv}{dt} = e^T[(A + GC)^T + (A + GC)]e - \frac{2\Delta\omega_r\left(e_{ids}\hat{\varphi}_{qr}^s - e_{iqs}\hat{\varphi}_{dr}^s\right)}{c} + \frac{2\Delta\omega_r}{\lambda}\frac{d\hat{\omega}_r}{dt} \tag{17}$$

Where,

$$e_{ids} = i_{ds} - \hat{\imath}_{ds} \text{ and } e_{iqs} = i_{qs} - \hat{\imath}_{qs}$$

By equalizing the second term with the third term, the following adaptation scheme is derived, i.e,

$$\frac{d\hat{\omega}_r}{dt} = \frac{\lambda}{c}(e_{ids}\hat{\varphi}_{qr}^s - e_{iqs}\hat{\varphi}_{dr}^s) \tag{18}$$

4. SIMULATION ANALYSIS AND RESULTS

An equivalent simulation model of the above estimation schemes is built in Simulink and the performance is observed for different values of load perturbations, incorrect parameter setting and Voltage Unbalance. The Induction motor is fed from a three phase ac voltage source of rating 415V, 50Hz and is run in the motoring mode. The model is run for two sets of load torque perturbations respectively:

   a) Step torque – Initially at no load, after a fixed time interval, it is increased to rated load of 200 Nm.
   b) For a Rated Load torque of 200 Nm, the effect of change in stator resistance and rotor time constant is observed in the performance of the estimators.
   c) For a Rated Load torque of 200 Nm, an unbalanced three phase voltage is supplied (each phase voltage having amplitude of 200 V, 180 V and 220 V respectively).

The motor ratings and the parameters considered for simulation are given as follows: A 50HP, three-phase, 415V, 50 Hz, star connected, four-pole induction motor with equivalent parameters: $R_s = 0.087\Omega$, $R_r = 0.228\Omega$, $L_{ls} = L_{lr} = 0.8$ mH, $L_m = 34.7$ mH, Inertia, $J = 1.662$ kgm^2, friction factor $= 0.1$.

4.1. Rotor Flux based MRAS Estimator
   a) Step Torque (Rated Load is applied at 5 seconds)

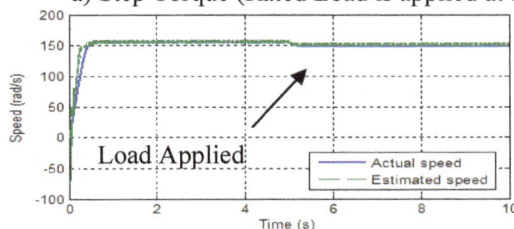

Figure 5. Speed tracking during step torque perturbation

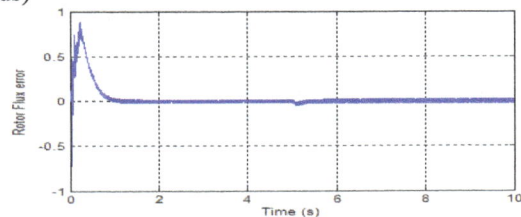

Figure 6. Rotor Flux error during step torque perturbation

b) Rated Torque (with incorrect setting of parameters)

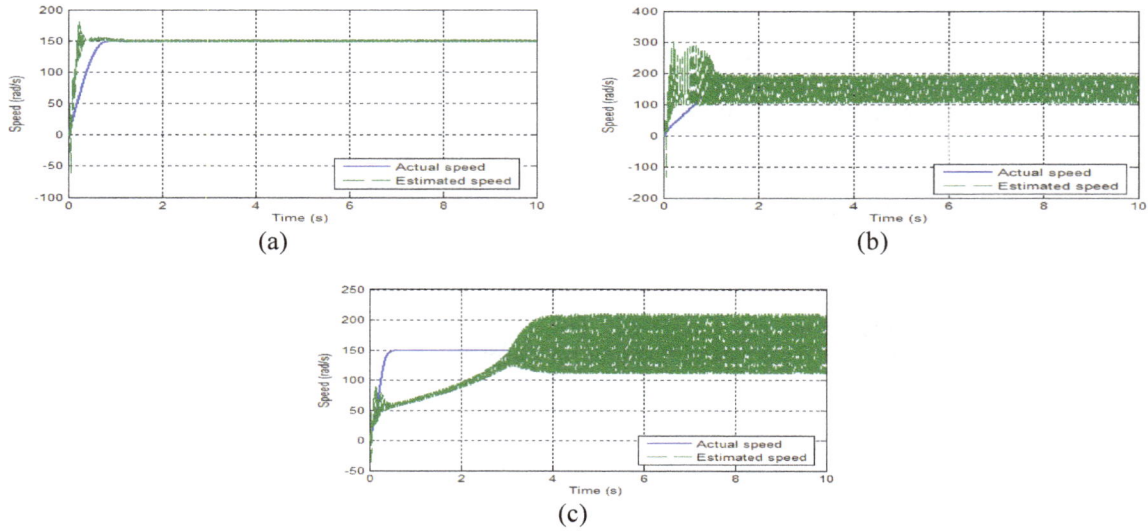

(a)

(b)

(c)

Figure 7. Speed tracking for different values of Stator Resistance (R_s) and Rotor Time constant (T_r)
(a) R_s, T_r (b) $0.5R_s$, $0.5T_r$ (c) $1.5R_s$, $1.5T_r$

c) Rated Torque (with Voltage Unbalance)

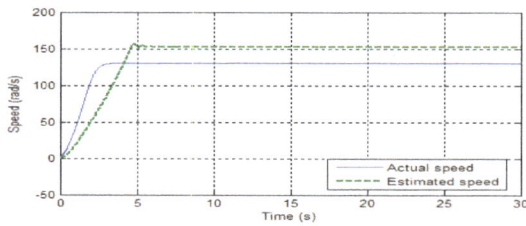

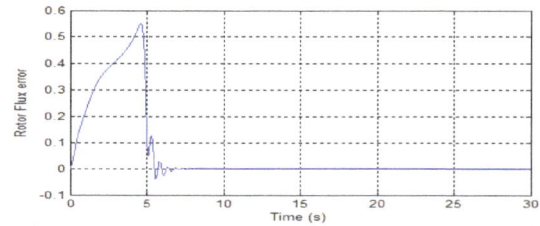

Figure 8. Speed tracking during Voltage Unbalance

Figure 9. Rotor Flux error during Voltage Unbalance

4.2. Adaptive Speed Observer
a) Step Torque (Rated Load is applied at 3.2 seconds)

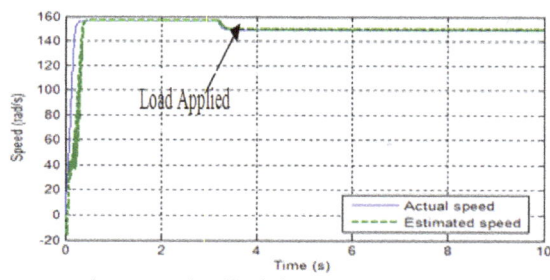

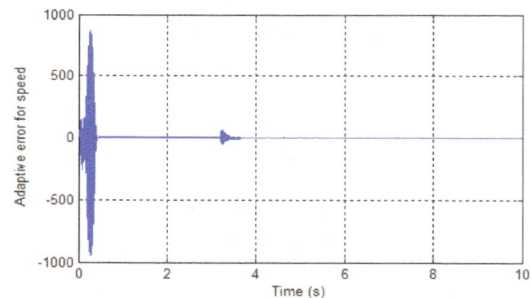

Figure 10. Speed tracking during step torque
perturbation

Figure 11. Adaptive error during step torque
perturbation

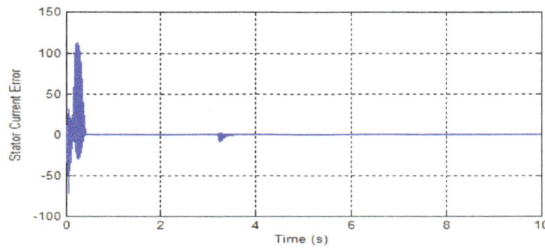

Figure 12. Stator Current error during step torque perturbation

b) Rated Torque (Incorrect setting of parameters)

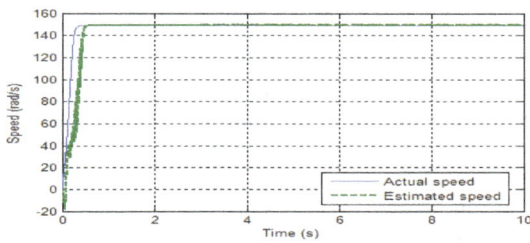

(a)

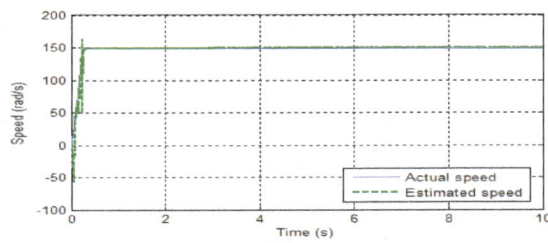

(b)

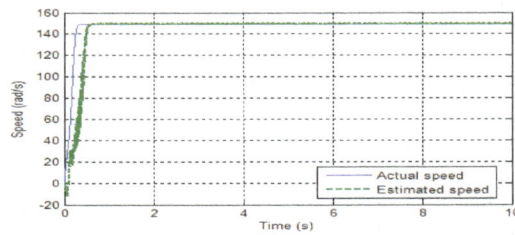

(c)

Figure 13. Speed tracking for different values of Stator Resistance (R_s) and Rotor Time constant (T_r)
(a) R_s, T_r (b) $0.5R_s$, $0.5T_r$ (c) $1.5R_s$, $1.5T_r$

c) Rated Torque (with Voltage Unbalance)

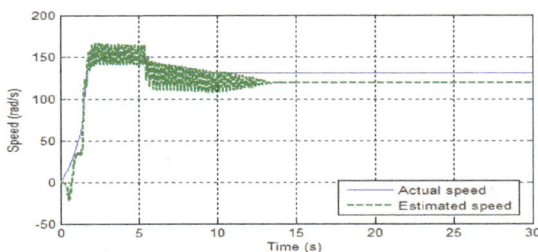

Figure 14. Speed tracking during Voltage Unbalance

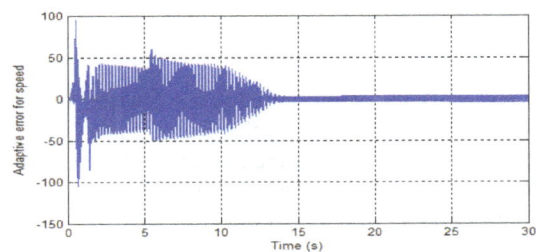

Figure 15. Adaptive speed error during Voltage Unbalance

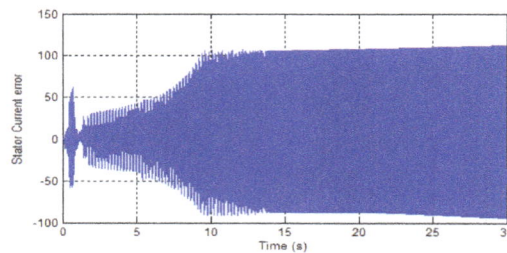

Figure 16. Stator Current error during Voltage Unbalance

The dynamic performance of the adaptive control schemes can be discussed based on the above results. In case of the Rotor Flux based MRAS Speed estimation scheme, it can be seen in Figure 5 that the estimated speed tracks the actual speed reasonably well even under step torque perturbations. The convergence of any adaption scheme is an important issue, in Figure 6 the Rotor flux error is seen to converge to zero, which is the reason the estimated speed tracks the measured speed in a very short time interval. Incorrect setting of parameters leads to high oscillations in the estimated speed which can be observed in Figure 7(b) and (c). It is also noticed that the estimated speed takes more time to track the actual speed when there is a 50% increase in the set value of the Stator resistance and Rotor time contant. During Voltage Unbalance, even though the Rotor flux error converges to zero (Figure 9), the estimated speed settles at a value of 152.9 rad/s compared to the actual speed which is about 130.2 rad/s (Figure 10). The difference in the values of the speeds can be attributed to the change in the flux level (both in the reference as well as the adjustable model) due to unbalance in supply voltage.

It can be distinctly seen that the Adaptive Speed Observer exhibits far superior tracking performance than the Rotor flux MRAS scheme. Though the tracking performance in Figure 10 is somewhat similar to that of the Rotor Flux MRAS, the difference lies when the same is subjected to parametric variations. For all the cases of Stator resistance and Rotor Time constant variations, a consistent, near smooth tracking performance is obtained which can be noticed in Figure 13(b) and (c). The adaptive error for speed derivation and the Stator current error for the observer gain converge exaclty to zero, which is a primary reason for the superior tracking performance. When an unbalanced voltage is supplied, initially there are oscillations in the estimated speed, but it settles at a value (120.5 rad/s) somewhat lower than the actual speed (130.8 rad/s) which can be noticed in Figure 14. But, the differences in the speeds is relatively lesser than that of the Rotor flux MRAS scheme. This can be pertaining to the high stator current error seen in Figure 16. which affects the correction term used for the adaptive observer model.

On comparing the performance of the above speed estimation schemes, the Adaptive Speed Observer is presented as a better alternative due to its robust tracking performance and reduced oscillations. It tracks the actual speed in a relatively less amount of time. This analysis confines itself to motoring mode at speed ranges from medium to base synchronous speed.

5. CONCLUSION

This paper presented a comparison of performance of two popularly used adaptive control based speed estimation schemes, the Rotor Flux MRAS and the Adaptive Speed observer, when subjected to variations in load torque, parameters and unbalanced supply voltage. It also presented the effect of the same on the convergence mechanisms of the above adaptive schemes. The Adaptive Speed observer is found to be more efficient and robust in tracking the actual speed. Though, it is also susceptible to speed errors, the scope can be extended for joint state estimation such that, there are no mismatch in parameters during low and very low speeds.

REFERENCES

[1]   Mohammad Jannati, *et al*. Speed Sensorless Vector Control of Unbalanced Three-Phase Induction Motor with Adaptive Sliding Mode Control, *International Journal of Power Electronics and Drive System (IJPEDS)*, 2014; 4(3): 406-418.

[2]   Mohamed S Zaky, A stable adaptive flux observer for a very low speed-sensorless induction motor drives insensitive to stator resistance variations, *A in Shams Engineering Journal*, 2011; 2: 11-20.

[3]   BK Bose, *Modern Power Electronics and AC Drives*, New Delhi, India: Prentice-Hall, 2006; 8: 350-440.

[4]   Teresa Orlowska-Kowalska, Mateusz Dybkowski, Stator – Current-Based MRAS Estimator for a Wide Range Speed-Sensorless Induction-Motor Drive, *IEEE Trans. Ind. Electron*, 2010; 57(4): 1296-1308.

[5]   Mohammad Jannati, *et al.*, Speed Sensorless Direct Rotor Field-Oriented Control of SinglePhase Induction Motor Using Extended Kalman Filter, *International Journal of Power Electronics and Drive System (IJPEDS)*, 2014; 4(4): 430-438.

[6]   H Madadi Kojabadi, *et al. Recent Progress in Sensorless Vector-Controlled Induction Motor Drives*, Proceedings of the Large Engineering Systems Conference on Power Engineering, IEEE 0-7803-7520-3, 2002; 80-85.

[7]   V Verma, *et al.* Performance of MRAS Based Speed Estimators for Grid Connected Doubly fed Induction Machines During Voltage Dips, *IEEE 978-1-4673-2729-9/12*, 2012; 1-8.

[8]   Ahmad Razani Haron, Nik Rumzi Nik Idris, *Simulation of MRAS-based Speed Sensorless Estimation of Induction Motor Drives using MATLAB/SIMULINK*, First International Power and Energy Conference, Putrajaya, Malaysia, 2006; 411-415.

[9]   H Kubota, K Matsuse, DSP-Based Speed Adaptive Flux Observer of Induction Motor, *IEEE Trans. Ind. Appl.*, 1993; 29(02): 344-348.

[10] Huang Zhiwu, *et al. Stability Analysis and Design of Adaptive Observer Based on Speed Sensorless Induction Motor*, Proceedings of the 26th Chinese Control Conference, 2007; 28-32.

[11] H Madadi Kojabadi, L Chang, Model Reference Adaptive System Pseudoreduced-Order Flux Observer for Very Low Speed and Zero Speed Estimation in Sensorless Induction Motor Drives, *IEEE 0-7803-7262-X/02*, 2002; 301-305.

[12] H Madadi Kojabadi, *et al.*, Stability Conditions of Adaptive Pseudo-Reduced-Order Flux Observer for Vector-Controlled Sensorless IM Drives, *IEEE CCECE-CCGEI*, Niagara Falls, 2004; 1313-1316.

19

A High Gain Observer Based Sensorless Nonlinear Control of Induction Machine

Benheniche Abdelhak[*] **Bensaker Bachir**[**]

[*]Département d'Electrotechnique, Université Badji Mokhtar, BP.12 Annaba, 23000, Algérie.
[**]Laboratoire des Systèmes Electromécaniques, Université Badji Mokhtar, BP.12, Annaba, 23000, Algérie.

	ABSTRACT
Keyword: Backstepping control High gain observer Induction machine Lyapunov stability	In this paper a sensorless Backstepping control scheme for rotor speed and flux control of induction motor drive is proposed. The most interesting feature of this technique is to deal with non-linearity of high-order system by using a virtual control variable to render the system simple. In this technique, the control outputs can be derived step by step through appropriate Lyapunov functions. A high gain observer is performed to estimate non available rotor speed and flux measurements to design the full control scheme of the considered induction motor drive. Simulation results are presented to validate the effectiveness of the proposed sensorless Backstepping control of the considered induction motor.

Corresponding Author:

Bachir Bensaker,
Laboratoire des Systèmes Electromécaniques,
Université Badji Mokhtar
BP.12, Annaba, 23000, Algérie.
E-mail: bensaker_bachir@yahoo.fr

1. INTRODUCTION

Induction motor (IM) compared to other types of electric machines, is used in a wide range of industrial applications. This is due to its excellent reliability, great robustness and less maintenance requirements. However, the induction motor model is complicated for various reasons, among them:
a) The dynamic behavior of the motor is described by a fifth-order highly coupled and nonlinear differential equations,
b) Rotor electric variables (fluxes and currents) are practically unmeasurable state variables,
c) Some physical parameters are time-varying (stator and mainly rotor resistance, due to heating, magnetizing induction due to saturation).

The first control schemes of induction motors were based on traditional scalar control that can guarantee only modest performance. In many applications, it is necessary to use more sophisticated controls as Field Oriented Control (FOC) proposed by Blaschke [1]. This type of control technique has led to a radical change in control of the induction machines. Thanks to the quality of dynamic performance that it brings. In the FOC technique, called also vector control, the torque and flux are decoupled by a suitable decoupling network. In this type of control technique, the flux and the torque components are controlled independently by the stator direct and quadratic currents respectively. Thus permits to control the induction motor (IM) as a separately excited DC motor [2]. The high performance of such strategy may be deteriorated in practice due to plant uncertainties.

Other techniques were conceived like input-output linearization technique that is based on the use of differential geometry theory to allow by a diffeomorphic transformation a state feedback control of the induction motor system [3]-[5]. This method cancels the nonlinear terms in the plant model and fails when the physical parameters are time-varying.

By contrast, the passivity based control doesn't cancel all the nonlinearity terms but ensure system stability, by adding a damping term to the total energy of the system. It is characterized by its robustness against the parameter uncertainties, however, its experimental implementation is still difficult [6], [7]. The sliding mode control is another control technique that is characterized by simplicity of design and attractive robustness properties. Its major drawback is the chattering phenomenon [8]-[10].

Since last two decades, the nonlinear control called "Backstepping" became one of the most popular control techniques for a wide range of nonlinear system classes [13]-[21]. It is distinguished by its ability to easily guarantee the global stabilization of system, even in the presence of parametric uncertainties [18]. The design of the control law is based mainly on the construction of appropriate Lyapunov functions. Its present form is due to Krstic, Anellakopoulos and Kokotovic [13] based on the Lyapunov stability tools, this approach offers great flexibility in the synthesis of the regulator and naturally leads itself to an adaptive extension case. This control technique offers good performance in both steady state and transient operations, even in the presence of parameter variations and load torque disturbances.

In order to implement a nonlinear sensorless control technique, to improve the robustness and the reliability of induction motor drives, it is necessary to synthesize a state observer for the estimation of non-measurable state variables of the machine system that are essential for control purposes [23]-[26]. Among the observation techniques one can use the high gain observer technique to design an appropriate sensorless control of IM drives.

In this paper a Backstepping control that involves non measurable state variables of the induction motor system is performed. In order to achieve a sensorless control a high gain observer is designed to estimate non measured state variable of the machine.

The paper is organized as follows: In section two the nonlinear induction motor model is presented. Backstepping speed and flux controllers design is presented in section three. The high gain observer technique is presented in the section four. In the fifth and final section simulation results and comment are presented.

2. INDUCTION MOTOR NONLINEAR MODEL

In order to reduce the complexity of the three phase induction motor model, an equivalent two phase representation is used under assumptions of linearity of the magnetic circuit and neglecting iron losses. This type of model is designed in the fixed stator reference frame (α, β).

In this paper, the considered induction motor model has stator current, rotor flux and rotor angular velocity as selected state variables. The control inputs are the stator voltage and load torque. The available induction motor stator current measurements are retained as the motor system outputs.

In these conditions, the nonlinear model of the induction motor can be expressed as the following"

$$\dot{x} = f(x) + g(x)u \tag{1}$$

$$y = h(x) \tag{2}$$

With,

$$f(x) = \begin{bmatrix} -\gamma i_{s\alpha} + \frac{K}{T_r}\varphi_{r\alpha} + p\omega\varphi_{r\beta} \\ -\gamma i_{s\beta} - p\omega K\varphi_{r\alpha} + \frac{K}{T_r}\varphi_{r\beta} \\ \frac{M}{T_r}i_{s\alpha} - \frac{1}{T_r}\varphi_{r\alpha} + p\omega\varphi_{r\beta} \\ \frac{M}{T_r}i_{s\beta} - p\omega\varphi_{r\alpha} - \frac{1}{T_r}\varphi_{r\beta} \\ \frac{pM}{jL_r}\left(\varphi_{r\alpha}i_{s\beta} - \varphi_{r\beta}i_{s\alpha}\right) - \frac{f}{j}\Omega - \frac{T_l}{j} \end{bmatrix} \tag{3}$$

$$g(x) = \begin{bmatrix} \frac{1}{\sigma L_s} & 0 & 0 & 0 \\ 0 & \frac{1}{\sigma L_s} & 0 & 0 \end{bmatrix}^T \tag{4}$$

And $\quad h(x) = \begin{bmatrix} 1 & 0 & 0 & 0 \\ 0 & 1 & 0 & 0 \end{bmatrix}$ $\qquad$ With: $\gamma = \frac{R_s}{\sigma L_s} + \frac{R_r M^2}{\sigma L_s L_r^2}, \sigma = 1 - \frac{M^2}{\sigma L_s L_r}, T_r = \frac{L_r}{R_r}$ and $K = \frac{M}{\sigma L_s L_r}$.

Where$x = [i_{s\alpha}, i_{s\beta}, \varphi_{r\alpha}, \varphi_{r\beta}, \omega]^T$ is the state vector and $u = [u_{s\alpha}, u_{s\beta}]^T$ the input vector control; with $i_{s\alpha}, i_{s\beta}$ as the stator currents and $\varphi_{r\alpha}, \varphi_{r\beta}$ as the rotor flux, $u_{s\alpha}, u_{s\beta}$ are the stator command voltages. Ω, R_r, L_r, R_s and L_s are the rotor angular velocity, the rotor resistance, the rotor inductance, the stator resistance and the stator inductance respectively.M is the mutual inductance between stator and rotor winding, p is the number of pair poles, j is the moment of inertia of the rotor, f is the viscous friction coefficient and T_l is the external load torque.

3. SPEED AND FLUX BACKSTEPPING CONTROLLER DESIGN

The Backstepping control design is based on the use of the so-called "virtual control" to systematically decompose a complex nonlinear control problem into simpler one, smaller ones, by dividing the control design into various design steps. In each step we deal with an easier, single-input single-output design problem, and each step provides a reference for the next design step. This approach is different from the conventional feedback linearization in that it can avoid cancellation of useful nonlinearities to achieve the stabilization and tracking objectives.

3.1. First Step

In the first step, it is necessary to specify the desired (reference) trajectories that the system must track, and design controllers to ensure good tracking error.

To this end, we define a reference trajectory,$y_{ref} = (\Omega_{ref}, \varphi^2_{ref})$, where Ω_{ref} and φ^2_{ref} are speed and rotor flux modul reference trajectories.

The speed tracking error $e_{1\Omega}$ and the flux magnitude tracking error $e_{1\varphi}$ are defined as:

$$e_{1\Omega} = \Omega_{ref} - \Omega \tag{5}$$

$$e_{1\varphi} = \varphi^2_{ref} - \varphi^2_r \tag{6}$$

With $\quad \varphi^2_r = \varphi^2_{r\alpha} + \varphi^2_{r\beta}$

The error dynamical equations are:

$$\dot{e}_{1\Omega} = \dot{\Omega}_{ref} - \left[\frac{pM}{jL_r} \left(\varphi_{r\alpha} i_{s\beta} - \varphi_{r\beta} i_{s\alpha} \right) - \frac{T_l}{j} - \frac{f}{j}\Omega \right] \tag{7}$$

$$\dot{e}_{1\varphi} = \frac{d}{dt} \left(\varphi^2_{ref} \right) - \left[\frac{2M}{T_r} \left(\varphi_{r\alpha} i_{s\alpha} + \varphi_{r\beta} i_{s\beta} \right) \right] + \frac{2}{T_r} \varphi^2_r \tag{8}$$

By setting the virtual control expressions below:

$$\alpha_1 = \left[\frac{pM}{jL_r} \left(\varphi_{r\alpha} i_{s\beta} - \varphi_{r\beta} i_{s\alpha} \right) \right] \tag{9}$$

$$\beta_1 = \left[\frac{2M}{T_r} \left(\varphi_{r\alpha} i_{s\alpha} + \varphi_{r\beta} i_{s\beta} \right) \right] \tag{10}$$

We can write (8) and (9) under the following form:

$$\dot{e}_{1\Omega} = \dot{\Omega}_{ref} - \alpha_1 + \frac{T_l}{j} + \frac{f}{j}\Omega \tag{11}$$

$$\dot{e}_{1\varphi} = \frac{d}{dt} \left(\varphi^2_{ref} \right) - \beta_1 + \frac{2}{T_r} \varphi^2_r \tag{12}$$

Let us check the tracking error dynamics stability by choosing the following candidate Lyapunov function:

$$v_1 = \frac{1}{2} \left[e^2_{1\Omega} + e^2_{1\varphi} \right] \tag{13}$$

The time derivative of (13) gives:

$$\dot{v}_1 = e_{1\Omega} \dot{e}_{1\Omega} + e_{1\varphi} \dot{e}_{1\varphi} \tag{14}$$

To render the time derivative of the Lyapunov function negative definite one has to choose the derivatives of the error tracking as follows:

$$\dot{e}_{1\Omega} = -c_1 e_{1\Omega} \tag{15}$$

$$\dot{e}_{1\varphi} = -d_1 e_{1\varphi} \tag{16}$$

In these conditions the virtual control, deduced form relations (11) and (12) become as the following:

$$\alpha_1 = c_1 e_{1\Omega} + \dot{\Omega}_{ref} + \frac{T_l}{j} + \frac{f}{j}\Omega \tag{17}$$

$$\beta_1 = d_1 e_{1\varphi} + \frac{d}{dt}\left(\varphi_{ref}^2\right) + \frac{2}{T_r}\left(\varphi_{ref}^2 - e_{1\varphi}\right) \tag{18}$$

Where c_1 and d_1 are the positive design gains that determine the dynamic of closed loop.

The time derivative of the candidate Lyapunov function is evidently negative definite, so the tracking error $e_{1\Omega}$ and $e_{1\varphi}$ can be stabilized.

3.2. Second Step

Previous references, chosen to ensure a stable dynamic of speed and flux tracking error, can't be imposed to the virtual controls without considering errors between them.

To this end, let us define the following errors:

$$e_{2\Omega} = \alpha_1 - \left[\frac{pM}{jL_r}\left(\varphi_{r\alpha} i_{s\beta} - \varphi_{r\beta} i_{s\alpha}\right)\right] \tag{19}$$

$$e_{2\varphi} = \beta_1 - \left[\frac{2M}{T_r}\left(\varphi_{r\alpha} i_{s\alpha} + \varphi_{r\beta} i_{s\beta}\right)\right] \tag{20}$$

One determines the new dynamics of the errors $e_{1\Omega}$ and $e_{1\varphi}$, expressed now in terms of $e_{2\Omega}$ and $e_{2\varphi}$.

$$\dot{e}_{1\Omega} = -c_1 e_{1\Omega} + e_{2\Omega} \tag{21}$$

$$\dot{e}_{1\varphi} = -d_1 e_{1\varphi} + e_{2\varphi} \tag{22}$$

From (19) and (20) we obtain the following errors dynamics equations:

$$\dot{e}_{2\Omega} = \alpha_2 + \left[\frac{pK}{j}\left(\varphi_{r\alpha} u_{s\beta} - \varphi_{r\beta} u_{s\alpha}\right)\right] \tag{23}$$

$$\dot{e}_{2\varphi} = \beta_2 - \left[2KR_r\left(\varphi_{r\alpha} u_{s\alpha} + \varphi_{r\beta} u_{s\beta}\right)\right] \tag{24}$$

Where

$$\alpha_2 = \dot{\alpha}_1 + \frac{pM}{jL_r}\left[\left(\gamma + \frac{1}{T_r}\right)\left(\varphi_{r\alpha} i_{s\beta} - \varphi_{r\beta} i_{s\alpha}\right)\right] + \frac{pM}{jL_r}\left[p\Omega[(\varphi_{r\alpha} i_{s\alpha} + \varphi_{r\beta} i_{s\beta}) + K\varphi_r^2]\right]$$

$$\beta_2 = \dot{\beta}_1 + \frac{2M}{T_r}\left[\left(\gamma + \frac{1}{T_r}\right)\left(\varphi_{r\alpha} i_{s\alpha} + \varphi_{r\beta} i_{s\beta}\right) - \frac{K}{T_r}\varphi_r^2\right] - \frac{2M}{T_r}\left[p\Omega(\varphi_{r\alpha} i_{s\beta} - \varphi_{r\beta} i_{s\alpha}) + \frac{M}{T_r}\left(i_{s\alpha}^2 + i_{s\beta}^2\right)\right]$$

One can see, from relations (23) and (24) that the real control components have appeared in the error dynamics. Thus permits us to construct the final Lyapunov function as:

$$v_2 = \frac{1}{2}\left[e_{1\Omega}^2 + e_{1\varphi}^2 + e_{2\Omega}^2 + e_{2\varphi}^2\right] \tag{25}$$

So the CLF derivative is determined below, by using (21), (22), (23) and (24):

$$\dot{v}_2 = -c_1 e_{1\Omega}^2 + e_{1\Omega} e_{2\Omega} - d_1 e_{1\varphi}^2 + e_{1\varphi} e_{2\varphi} - c_2 e_{2\Omega}^2 - d_2 e_{2\varphi}^2 + e_{2\Omega}\left(c_2 e_{2\Omega} + \alpha_2 + \frac{pK}{j}\left(\varphi_{r\alpha} u_{s\beta} - \right.\right.$$

$$\varphi_{r\beta} u_{s\alpha} + e_{2\varphi} d_1 e_{1\varphi} + \beta_2 - 2KR_r 2KR_r \varphi_{r\alpha} u_{s\alpha} + \varphi_{r\beta} u_{s\beta} \tag{26}$$

Where c_2 and d_2 are the positive design gains that determine the dynamic of closed loop.

In order to make the CLF derivative negative definite as:

$$\dot{v}_2 = -c_1 e_{1\Omega}^2 - d_1 e_{1\varphi}^2 - c_2 e_{2\Omega}^2 - d_2 e_{2\varphi}^2 \leq 0 \tag{27}$$

We choose voltage control as follows:

$$c_2 e_{2\Omega} + \alpha_2 + \frac{pK}{j}\left(\varphi_{r\alpha} u_{s\beta} - \varphi_{r\beta} u_{s\alpha}\right) = 0 \tag{28}$$

$$d_1 e_{1\varphi} + \beta_2 - 2KR_r\left[2KR_r\left(\varphi_{r\alpha} u_{s\alpha} + \varphi_{r\beta} u_{s\beta}\right)\right] = 0 \tag{29}$$

Thus leads to the following control expressions:

$$u_{s\alpha} = \frac{1}{\varphi_r^2}\left[\frac{(\beta_2 + e_{2\varphi} + d_2 e_{2\varphi})}{2KR_r}\varphi_{r\alpha} - \frac{j}{pK}\left[\alpha_2 + e_{1\varphi} + c_2 e_{2\varphi}\right]\varphi_{r\beta}\right] \tag{30}$$

$$u_{s\beta} = \frac{1}{\varphi_r^2}\left[\frac{(\beta_2 + e_{2\varphi} + d_2 e_{2\varphi})}{2KR_r}\varphi_{r\beta} + \frac{j}{pK}\left[\alpha_2 + e_{1\varphi} + c_2 e_{2\varphi}\right]\varphi_{r\alpha}\right] \tag{31}$$

4. NONLINEAR HIGH GAIN OBSERVER DESIGN

Generally, the dynamic behavior of the induction motor (IM) belongs to a class of relatively fast systems. For computational issue, the high gain observer which admits an explicit correction gain can be considered as one of the most viable candidate in the problem of state estimation. Later on, we adopt this method in our design.

Consider the nonlinear uniformly observable class of systems as the following form [23].

$$x = f(x, u) + \varepsilon \tag{32}$$

$$y = Cx = x^1 \tag{33}$$

Where the state $x \in \mathbb{R}^n$ with $x^k \in \mathbb{R}^p$ for $k = 1, 2, \ldots, q$ and $n_1 \geq n_2 \geq \cdots n_q$. The input $\subset \mathbb{U}$ a compact set of $\mathbb{R}^m$, the output $y \in \mathbb{R}^{n_1}$.

$$x = \begin{bmatrix} x^1 \\ x^2 \\ \vdots \\ x^q \end{bmatrix}; f(x, u) = \begin{bmatrix} f_1(x^1, x^2, u) \\ f_2(x^1, x^2, x^3, u) \\ \vdots \\ f_{q-1}(x^1, x^2, \ldots, x^{q-1}, u) \\ f_q(x, u) \end{bmatrix}; \varepsilon = \begin{bmatrix} 0 \\ \vdots \\ 0 \\ \varepsilon^{q-1} \\ \varepsilon^q \end{bmatrix}$$

With I_{n_1} is the $n_1 \times n_1$ identity matrix and $0_{n_1 \times n_j}$ is the $n_1 \times n_l$ null matrix, $l \in \{2, \ldots, q-1\}$. $\varepsilon^k \in \mathbb{R}^{n_k}, k \in \{q-1, q\}$, each ε^k is an unknown bounded real valued function that depend on uncertain parameters, in our case we propose $\varepsilon^k = 0$.

The synthesis of the high gain observer (HGO) corresponding to systems of the form (32) and (33), requires making some assumptions as follows:

a) There exist γ, δ with $0 < \gamma \leq \delta$ such that for all $k \in \{1, \ldots, q-1\}, x \in \mathbb{R}^n, u \in U$ we have:

$$0 < \gamma^2 I_{nk} \leq \left[\frac{\partial f_k(x^{1:k}, u)}{\partial x^{k+1}}\right]^T \frac{\partial f_k(x^{1:k}, u)}{\partial x^{k+1}} \leq \delta^2 I_{nk}$$

Moreover, we assume that $Rank\left(\frac{\partial f_k(x^{1:k}, u)}{\partial x^{k+1}}\right) = n_{k+1}$

b) The function $f(x, u)$ is globally Lipchitz with respect to x, uniformly in u.

In these conditions the high gain observer corresponding to systems of the form (32) and (33) can be written as:

$$\dot{\hat{x}} = f(\hat{x}, u) - \theta\Lambda^{-1}(\hat{x})\Delta_\theta^{-1}S^{-1}C^T\bar{C}(\hat{x} - x) \tag{34}$$

Where $\Lambda^{-1}(\hat{x})$ is the left inverse of block diagonal matrix $\Lambda(\hat{x})$ defined as:

$$\Lambda(\hat{x}) = blockdiag\left[I_{nk}, \frac{\partial f_k(x^{1:k},u)}{\partial x^{k+1}}, \dots, \prod_{i=1}^{q-1}\frac{\partial f_k(\hat{x},u)}{\partial \hat{x}^{k+1}}\right]$$

$$\Delta_\theta(\hat{x}) = blockdiag\left(I_{n_1}, \frac{1}{\theta}I_{n_1}, \dots, \frac{1}{\theta^{q-1}}I_{n_1}\right), \ \theta > 0$$

θ is a real number representing the only design parameter of the observer.

S is a definite positive matrix, solution of the following algebraic Lyapunov equation:

$$S + A^T S + SA = C^T C \tag{35}$$

With $C = [I_{n_1}, 0_{n_1}, \dots, 0_{n_1}]$ and $\mathcal{A} = \begin{bmatrix} 0 & \bar{\mathcal{A}} \\ 0 & 0 \end{bmatrix}$, with:

$$\bar{\mathcal{A}} = blockdiag(I_{n_1}, 0_{n_1}, \dots, 0_{n_1}) \in \mathbb{R}^{n_1(q-1)}$$

Note that relation (35) is independent of the system parameters and the solution can be expressed analytically. For a straightforward computation, its stationary solution is given by:

$$S(i,j) = (-1)^{i+j}C_{i+j-2}^{j-1}I_{n1} \tag{36}$$

Where $C_j^i = \frac{j!}{i!(j-i)!} for \ 1 \le i,j \le q$.

In these conditions we can explicitly determinate the correction gain of (34) as follows:

$$\theta\Lambda^{-1}(\hat{x})\Delta_\theta^{-1}S^{-1}C^T = \begin{bmatrix} \theta C_1^q I_{n1} \\ \theta^2 C_2^q \left[\frac{\partial f_1}{\partial x^2}(x,u)\right]^- \\ \vdots \\ \theta^q C_q^q \left[\prod_{i=1}^{q-1}\frac{\partial f_k}{\partial x^{k+1}}(x,u)\right]^- \end{bmatrix} \tag{39}$$

It should be emphasized that the implementation of HGO is quite simple.

5. SIMULATION RESULTS AND COMMENTS

To investigate the usefulness of the proposed sensorless control approach a simulation experiments have been performed for a three-phase induction motor, whose parameters are depicted in Table 1.

Table 1. Induction motor parameters

Symbol	Quantity	N. Values
P_a	Power	0.75KW
F	Supply frequency	50HZ
P	Number of pair poles	2
V	Supply voltage	220V
R_s	Stator resistance	10Ω
R_r	Rotor resistance	6.3Ω
L_s	Stator inductance	0.4642H
L_r	Rotor inductance	0.4612H
L_m	Mutual inductance	0.4212H
ω_r	Rotor angular velocity	157rd/s
J	Inertia coefficient	$0.02Kg^2/s$
f	Friction coefficient	0N.s/rd

Two schemes of high gain state observer for the estimation of IM states are investigated. The first scheme is dedicated to electromagnetic state variables estimation of the considered induction motor, while the second scheme performs the estimation of mechanical state variables namely the rotor speed and the load torque.

Based upon the estimated and measured state variables, Backstepping controllers of the rotor speed and rotor flux are respectively implemented using Matlab/Simulink software programming. The obtained simulation results are presented in Figure 1 to Figure 5.

From Figure 1 to Figure 5 (Figure 1-5) the reference, measured and estimated state variables of the machine are presented according to load torque variation from no load value to the value $T_l = 5N.m$, introduced between [0.5s-1.5s]. This simulation is carried out by applying a reference speed as illustrated in Figure 1. The measured and estimated speed converges perfectly to their reference. One can see also that a significant decoupling effect of flux components under rotor angular speed and load torque variations. Figure 2 shows that measured and estimated rotor flux tracks the reference flux with no disturbance are found, Figure 3 note that the proposed approach exhibits high accuracy in torque tracking when the reference torque change, figures (4-5) show the measured and estimated stator currents and rotor flux components respectively. Analysis of the simulation results shows that the obtained performance of rotor angular speed and flux tracking are very adequate. Analysis of the different figures points out that designed nonlinear observer (High gain) effectively estimates the unmeasured state variables of the machine and tracks the load torque variations with respect to applied nonlinear control law computed in accordance with the Backstepping control technique.

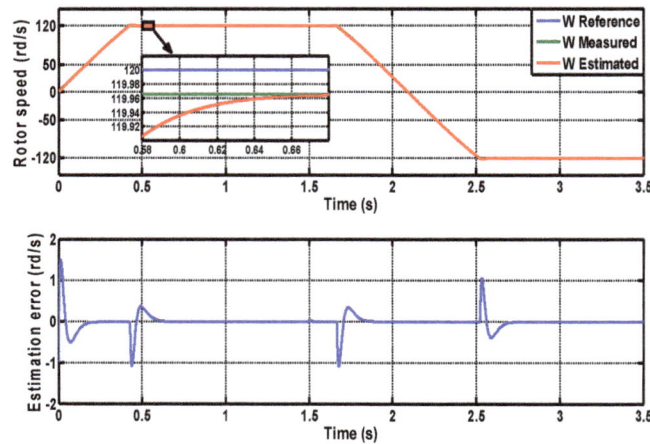

Figure 1. Reference measured and estimated Rotor speed evolution according to load variations

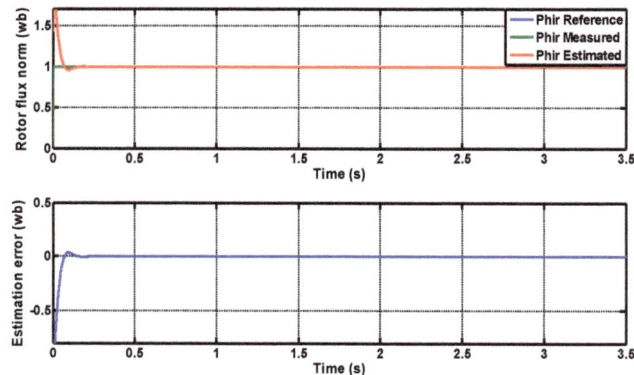

Figure 2. Reference measured and estimated norm of the rotor flux and estimation error

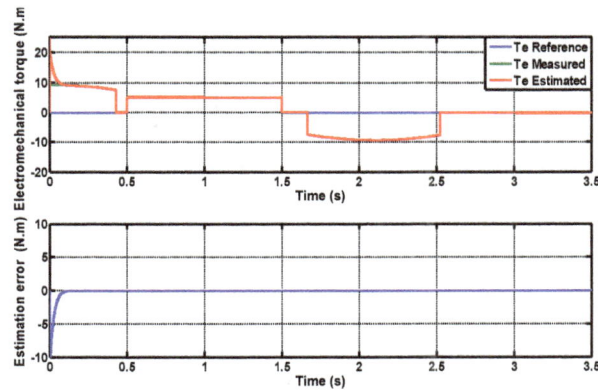

Figure 3. Reference measured and estimated electromechanical torque

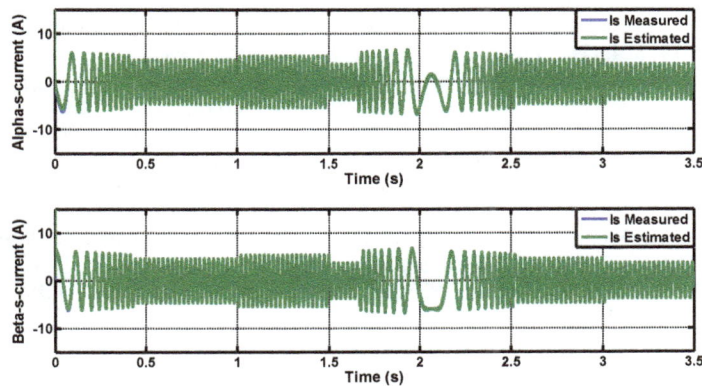

Figure 4. Measured and estimated $(\alpha - \beta)$ stator currents

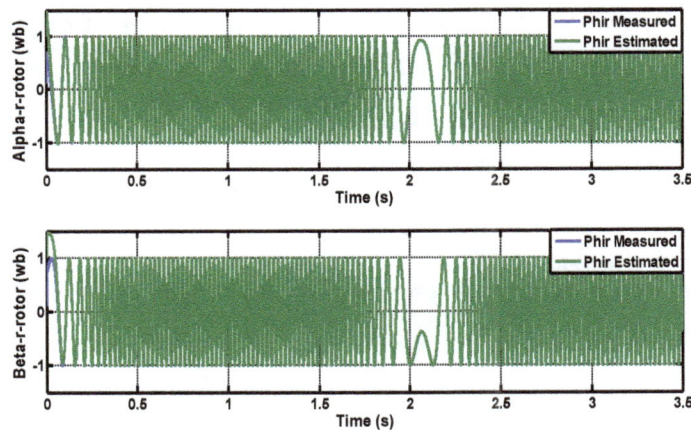

Figure 5. Measured and estimated $(\alpha - \beta)$ rotor flux

6. CONCLUSION

In this paper, we have investigate the possibility to implement a sensorless speed control of the induction machine using the technique of Backstepping, associated with a speed observer based on the high gain approach. The simulation results showed that this approach of control presents good performances and allows a complete decoupling between the flux and the torque. The machine keeps these performances. This technique can be improved further by using online estimation of parameters. On the other hand, simulation results show that this approach improves the performance of trajectory tracking and should bypass shortcomings of conventional methods. To this end, experimental tests will be investigated in a future framework.

REFERENCES

[1] F Blaschke. The Principle of Field Orientation Applied to the Transvector Closed-loop Control System for Rotating Field Machines. *Siemens Rev.* 1972; 34(5): 217-220.

[2] D Casadei, G Grandi, G Serra. Rotor Flux Oriented Torque-Control of Induction Machines Based on Stator Flux Vector Control. *Proceedings of the EPE Conference*, Brighton. 1993; 5: 67-72.

[3] R Marino, S Peresada, P Valigi. Adaptive Input-Output Linearizing Control of Induction Motors. *IEEE Transactions on Automatic Control.* 1993; 38(2): 208- 221.

[4] M Bodson, J Chiasson, R Novotnak. High-Performance Induction Motor Control Via Input-Output Linearization. *IEEE Control Syst Mag.* 1994; 14(4): 25-33.

[5] M Moutchou, A Abbou, H Mahmoudi, M Akherraz. Sensorless Input-Output Linearization Speed Control of Induction Machine. *The international workshop on Information Technologies and Communication (WOTIC'11)*, ID.123, 2011.

[6] R Ortega, G Espinoza. Torque Regulation of Induction Motor. *Automatica.* 1993; 29(3): 621-633.

[7] R Ortega, *et al.* On Speed Control of Induction Motor. *Automatica.* 1996; 32(3): 455-460.

[8] VI Utkin. Sliding Mode Control Design Principles and Applications to electric drives. *IEEE Transactions on Industrial Electronics.* 1993; 40(1): 23-36.

[9] A Sabanovic, DB Izosimov. Application of Sliding Modes to Induction Motor Control. *IEEE Transactions on Industry Applications.* 1981; 17: 344-348.

[10] G Bartolini, A Ferrara et E Usai. Chattering Avoidance by Second Order Sliding Mode control. *IEEE transactions on Automatic Control.* 1998; 43(2): 241-246.

[11] A Gouichiche, MS Boucherit, A Safa, Y Messlem. Sensorless Sliding Mode Vector Control of Induction Motor Drives. *International Journal of Power Electronics and Drive System (IJPEDS).* 2012; 2(3): 277-284.

[12] O Boughazi, A Boumedienne, H Glaoui. Sliding Mode Backstepping Control of Induction Motor. *International Journal of Power Electronics and Drive System (IJPEDS).* 2014; 4(4): 481~488.

[13] M Krstic, I Kannellakopoulos, P Kokotovic. Nonlinear and Adaptive Control Design. *Wiley and Sons Inc*, New York, 1995.

[14] R Trabelsi, A Khedher, MF Mimouni, F M'sahlic. Backstepping Control for an Induction Motor Using an Adaptive Sliding Rotor-Flux Observer. *Electric Power Systems Research.* 2012; 93: 1-15.

[15] F Ikhouane, M Krstic. Adaptive BacksteppingWith arameter Projection: Robustness and Asymptotic Performance. *Automatica.* 1998; 34: 429-435.

[16] RA Freeman, PV Kokotovic. Backstepping Design of Robust Controllers for a Class of Nonlinear Systems. *Proceedings of IFAC Nonlinear Control Systems Design Symposium, Bordeaux France.* 1992; 307-312.

[17] HTan, JChang. *Adaptive Backstepping Control of Induction Motor with Uncertainties.* Proceedings of the American control conference, San Diego, California. 1999.

[18] HTLee, LC Fu, FL Lian. Sensorless Adaptive Backstepping Speed Control of Induction Motor. Proc IEEE Conference on Decision Control, USA. 2002: 1252-1257.

[19] M Ghanes. Tracking Performances of Backstepping and High Gain Observers for Sensorless Induction Motor Control Against low Frequencies Benchmark. *IEEE International Conference on Control Applications, CCA.* 2007; 652–657.

[20] R Trabelsi, A Khedher, MF Mimouni, FM'sahlic. Backstepping Control for an Induction Motor with an Adaptive Backstepping Rotor Flux Observer. *18th Mediterranean Conference on Control & Automation, Marrakech, Morocco.* 2010.

[21] A Ebrahim, G Murphy. Adaptive Backstepping Control of an Induction Motor under Time-Varying Load Torque and Rotor Resistance Uncertainty. *Proceedings of the 38th Southeastern Symposium on System Theory, Tennessee Technological University Cookeville,* TN, USA. 2006.

[22] M Moutchou, A Abbou, H Mahmoudi. Sensorless Speed Backstepping Control of Induction Machine, Based On Speed MRAS Observer. *(ICMCS). International Conference on Multimedia Computing and Systems, Tangier, Morocco.*2012.

[23] S Hadj-Said, MF Mimouni, FM Salhi, M Farza. High Gain Observer Based on-line Rotor and Stator Resistance Estimation for IMs. *Simulation practice and Theory.* 2011; 19: 1518-1529.

[24] A Abbou, H Mahmoudi, A Elbacha. High-Gain Observer Compensator for Rotor Resistance Variation On Induction Motor RFOC. *14th IEEE International Conference on Electronics, Circuits and Systems, ICECS, Marrakech, Morocco.* 2007.

[25] A Dib, M Farza, M M'Saad, Ph Dorleans, JF Massieu. High Gain Observer for Sensorless Induction Motor. *Preprints of the 18th IFAC World Congress Milano (Italy).* 2011.

[26] G Bornard, H Hamrnouri. A high gain observer for a class of uniformily observable Systems. *IEEE, Brignton, Great Britain,* 1991.

20

Versions of Switched Reluctance Generator Design at a Constant Stator Configuration

N.V. Grebennikov*, A.V. Kireev**,
* Rostov State Transport University, Rostov-on-Don, Russia
** Science and Technology Center "PRIVOD-N", Rostov-on-Don, Russia

ABSTRACT

Keyword:

Computer model
Number of phases
Overlap factor
Switched reluctance machines

The investigation of the influence of the number of phases of switched reluctance generator (SRG) to the pulse of electromagnetic torque was carried out. The computer model was created. The amplitude of torque ripples reduces to 6 times with increasing of the ripple frequency to 5 times, that is more acceptable in terms of requirements.

Corresponding Author:

N.V. Grebennikov,
Rostov State Transport University,
NarodnogoOpolcheniya sq., Rostov-on-Don, Russia, Postal code: 344038.
Email: grebennikovnv@mail.ru

1. INTRODUCTION

Switched reluctance machines (SRM), designed as a high efficiency type of electromechanical energy converter [1]-[5], can be applied on vehicles including railway rolling stocks.

Electrical machines used on vehicles operate in severe conditions. During operation they are affected by significant dynamic forces resulting from vibration and shock particularly at high running speeds. It can cause to various failures: wires and winding connection disruption, cracking and insulating materials damage. For this reason when choosing electrical machines design there is a tendency to use simple and reliable technical solutions.

From this point of view the main advantage of SRM is the design simplicity. The rotor is passive without winding and the stator is equipped with winding consisting of centered type coils. In comparison with other types of electrical machines, SRM is more sophisticated, has less specific consumption of cooper and insulating materials. In case of SRM application on the vehicles, it will allow to improve the reliability of energy supply system, to achieve better energy and weight-size parameters, to reduce the cost and operation expenses. The disadvantages of SRM are considerable electromagnetic torque ripple and higher noise level.

2. VERSIONS OF SWITCHED RELUCTANCE GENERATOR

Consider the possibility to reduce the torque ripple on the example of switched reluctance generator (SRG) having classical configuration 18/12 (18 teeth at the stator and 12 - at the rotor). 18 coils located at the stator are divided into three phases and the angle between coils' axles is 60°.

In [6] it is said that «if the terms of SRM should ensure the high stability of rotation frequency and low torque ripple, the number of phases should be chosen as maximum possible». Increasing the number of

SRM phases can be obtained by changing the number of teeth at the rotor while maintaining the same stator with 18 teeth, it is specified by economic considerations.

Consider the possible variants of SRG configuration, with stator having 18 teeth and coils disposed at each tooth, depending on number of teeth at the rotor (the number of teeth at the rotor is less than or equal to 18):

a) 18/18 – a single phase machine, with strong torque;

b) 18/9 – two-phase machine, which torque ripple is higher than three-phase machine obtains [6], with account of the fact that this ripple is conceptually impossible to eliminate [7];

c) 18/12 – three-phase machine, the number of stator teeth was increased threefold compared to basic three-phase machine configuration 6/4. It allows to reduce the noise level [6];

d) 18/16 – nine-phase machine with alternating polarity of adjacent windings and strong mutual inductance of adjacent phases. Note that the number of power semiconductor devices (PSD) in the converter is increased by three times compared to the converter of three-phase machine.

Based on analysis of SRG performance with different numbers of rotor teeth, six-phase machine with 18/15 configuration is proposed for application. The cost of the converter for this kind machine will be much lower than for nine-phase machine: the design of converter is known for machine's supply (Figure 1) having the same numbers of PSD as well as for three-phase machine [8].

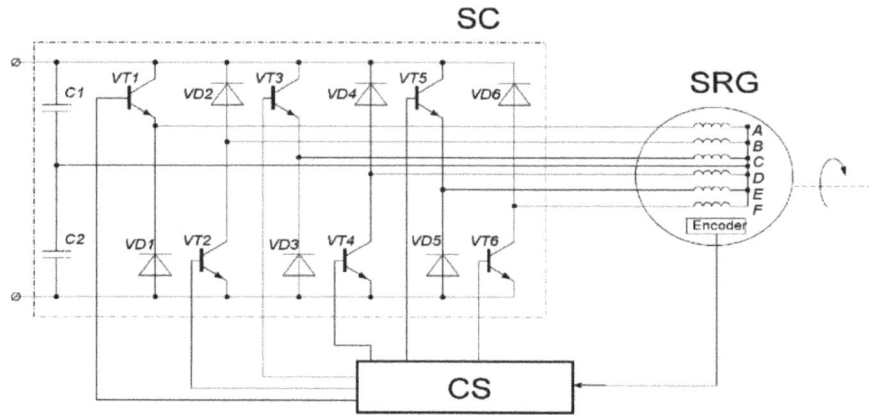

Figure 1. Scheme of six-phase SRG and supply converter (SC – Static converter, CS – control system, Encoder – rotor position sensor)

The main condition for elimination of electromagnetic torque ripple in SRM is partial overlapping of machine's operation areas by adjacent phases. To estimate the possible overlapping of these areas, two factors are used [7], [8]:

1) absolute overlapping factor:

$$\rho_A = m / 2,$$
where   m – number of SRM phases;

2) effective overlapping factor:

$$\rho_E = \frac{N_R}{2(N_S - N_R)},$$
where N_R – number of rotor teeth,

N_S – number of stator teeth.

From the formula analysis given above, it follows that the increase of absolute overlapping factor is possible under condition of increasing the number of phases, and the increase of effective overlapping factor is provided with increasing the number of rotor teeth.

Versions of SRM configuration and the values of overlapping factors are given in Table 1. Table 1 shows that the minimal torque ripples take place when maximum possible value of phase number, but it increases significantly the quantity of PSD and the cost of converter becomes the highest. From the other hand, the lowest cost of the converter is achieved at the lowest possible value of phase number (a single phase), but this variant is not rational for vibroacoustic indicators. The low torque ripple of SRG and the cost of its control system are mutually conflicting criteria, so the proposed variant of 18/15 configuration is optimal [9].

Table 1. Versions of SRM configuration

Teeth at stator	Teeth at rotor	Number of phases	Number of coils in the single phase	Number of PSD	Absolute overlapping factor, ρ_A	Effective overlapping factor, ρ_E
18	18	1	18	4	0,5	–
18	9	2	9	8	1	0,5
18	12	3	6	12	1,5	1
18	15	6	3	24 (12)	3	2,5
18	16	9	2	36	4,5	4

3. COMPUTER SIMULATION

To investigate different operation modes of SRG and to develop the optimal control of phase switching, the computer simulator of SRG electrical part has been developed in software package Matlab/Simulink (Figure 2), with account of mutual phase inductance.

The management of keys VT1…VT6 switching is performed by unit Upravlenie, based on received signals: w(rad/s) – angular frequency of rotor rotation, pos – angular rotor position relative to stator, I(A) – current value in SRG windings.

The initial supply impulse goes to generator from pre-charged capacitor C.

Units Scope w, Scope I(A)_V(V), Scope Moment, Scope Flux(V*s), Scope V_n and Scope I_n are designed for oscillograms recording of the following: angular rotation frequency of current and voltage in stator windings of generator, electromagnetic torque, flux linkage, load voltage and load current.

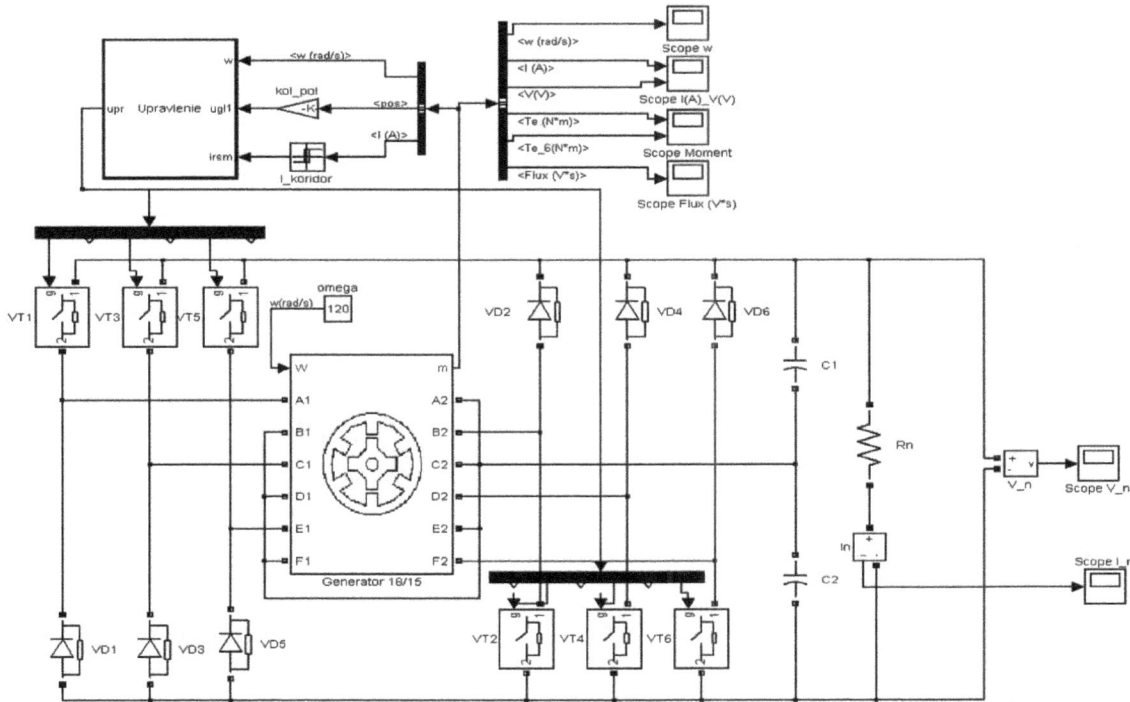

Figure 2. Computer model of six-phase SRG 18/15 configuration

As a result of simulation it was obtained current, voltage in SRG windings and electromagnetic torque dependences on angular rotor position for three-phase and six-phase configuration of SRG.

Figure 3 presents the diagrams of phase torques (curves 1 and 2) and total electromagnetic torques (curves 3 and 4) in relative units.

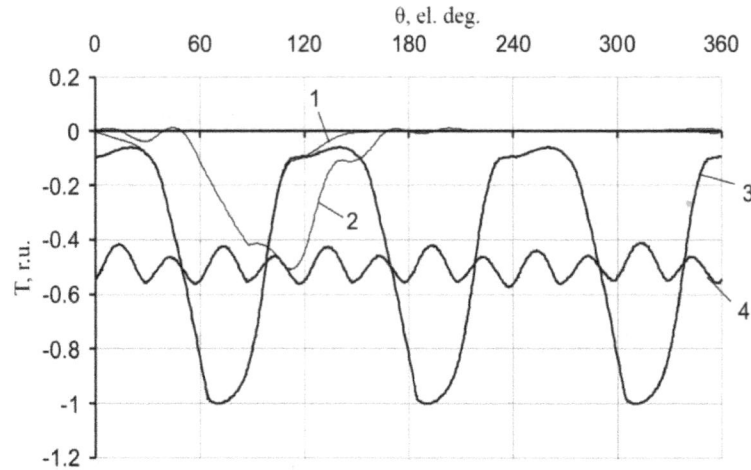

(1 – torque of single phase of SRG 18/12, 2 – torque of single phase of SRG 18/15, 3 – total torque of SRG 18/12 , 4 – total torque of SRG 18/15)

Figure 3. Dependence of SRG electromagnetic torques

The comparison of electromagnetic torques is given in Table 2 which demonstrates that the frequency of electromagnetic torque ripples at six-phase SRG configuration increases by about five times and amplitude of torque ripples decreases approximately by six times.

Table 2. Comparison of electromagnetic torques of SRG different configuration

SRG configuration	Number of SRG phase	Rotor rotation frequencyω_p, rad/s	Torque ripple frequency , Hz	Minimal torque value M_{MIN}, r.u	Ripple amplitude M_A, r.u.	$M_A/ |M_{MIN}|$
18/12	3	100	573	−1	0,469	0,469
18/15	6	100	2860	−0,57	0,081	0,141

4. CONCLUSION

Replacing the rotor having 12 teeth with the rotor having 15 teeth while maintaining the same stator and power converter allows us to reduce electromagnetic torque ripple approximately by 6 times.

ACKNOWLEDGEMENTS

The presented work has been developed with support of Russian Ministry of Education, grant RFMEFI57614X0036.

REFERENCES

[1] Voron OA, Grebennikov NV, Zarifian AA, Petrushin AD. Undercar switched reluctance generator. *Bulletin of the Russian Research and Design Institute for electric.* 2009; 1: 132-143.
[2] Switched reluctance electric drive for electric rolling stock. *Bulletin of the Russian Research and Design Institute for electric.* 2002; 44: 336.
[3] Kireev AV. Electrorolling composition with traction switched reluctance electric drive. *Rolling stock of the XXI century.* 2008; 20-22.

[4] Maged NF Nashed, Samia M Mahmoud, Mohsen Z El-Sherif, Emad S Abdel-Aliem. Hysteresis Current Control of Switched Reluctance Motor in Aircraft Applications. *International Journal of Power Electronics and Drive System.* 2014; 4(3): 376-392.

[5] Srinivas P, Prasad PVN. Direct Instantaneous Torque Control of 4 Phase 8/6 Switched Reluctance Motor. *International Journal of Power Electronics and Drive System.* 2011; 1(2): 121-128.

[6] Kuznetsov VA, Kuz'michevVA. Switched reluctance motor, 2003; 68 pp.

[7] Krasovskiy AB, Bychkov MG. Investigation of torque ripple in the switched reluctance electric drive. *Electricity.* 2001; 10: 33-43.

[8] Miller TGE. Switched Reluctance Motors and Their Control. *Oxford Magna Physics Publishing and Clarendon Press.* 1993.

[9] Grebennikov NV, Petrushin AD, Reactive switched electrical machine with rotation symmetr, *Rostov-na-Donu,* 2012; 13.

21

Design and Analysis of Drive System with Slip Ring Induction Motor for Electric Traction in India

C. Nagamani*, R. Somanatham**, U. Chaitanya Kumar***

* Research Scholar, University College of Engineering, Osmania University, Hyderabad, India
** HOD, Dept. Of Electrical & Electronics Engineering, Anurag College of Engineering, Hyderabad, India
*** M.Tech Student, Dept. Of EEE, Anurag College of Engineering, Hyderabad, India

Keyword: Electric Traction Slip Power Recovery Squirrel Cage Induction Motor Wound Rotor Induction Motor	**ABSTRACT** The use of Squirrel Cage Motor for Traction has revolutionised the motive power of a Locomotive. The Asynchronous Motor is rugged, has high starting Torque, very smooth Voltage and Speed control as compared to a DC Series Motor. When looking at the Traction perspective, a Wound Rotor Induction Motor can be an alternative to the Squirrel Cage Motor as it has higher starting Torque at lower starting current and better efficiency than a Squirrel Cage Motor. The Slip Power Recovery scheme also plays a proactive role as there can be substantial savings of energy in case of a Wound Rotor Induction Motor as the Slip Power recovered can be used to drive the Auxiliary Loads of the Locomotive and also for powering the trailing Passenger Cars. A detailed design and analysis of a Drive System with Wound Rotor Induction Motor for Electric Traction is presented in this Research Paper.

Corresponding Author:

C. Nagamani,
Department of Electrical Engineering,
Osmania University, Hyderabad, India,
Flat No.1, 12-12-173, Sridevi Apartments, Sitaphalmandi, Secunderabad 500061, India.
Email: cnagamani2025@gmail.com

1. INTRODUCTION

The invention of Gate Turn-Off Thyristors revolutionised the modern Electric Traction. Considerable advancements have taken place from the use of Diode Rectifiers for series-parallel control of the DC Series Motors to the use of simple Pulse-Width Modulation Technique for the control of Asynchronous Motors. The Inverter driven Induction Motor maintain a low-slip operation even during starting [1]. As a result, the Three-phase Squirrel Cage Induction Motor became very popular as the Traction Motor because of its properties of ruggedness, high starting torque, easy Motor Control through Micro-processor, regeneration up to zero speeds, efficiency of operation and better adhesion provided by it in preventing Wheel Slips. It is also observed that a WRIM can produce more starting torque as compared to a SQIM at lesser starting current. Also the Braking capability of WRIM has been found superior to SQIM [2]. With the ever increasing loads to be hauled, multiple operations of Locomotives have become the norm of the day. Under these conditions, the production units manufacturing Locomotives are working out ways and means to increase the Power of the Locomotives. In near future one might find Locomotives of the capacity of 7000 HP hauling heavy loads over various gradients in Indian Railways. When it comes to ratings of Motors for this kind of ultra high power Locomotives, the Motors could be rated at order of 900 kW to 1 MW.

In this given scenario, if one can harness the Power wasted due to 'Slip', the energy savings can be substantial. It has been proved in the case of high capacity Roller Mills, that the high initial cost of a Slip ring Induction Motor is overcome by the Slip Power Recovery Scheme implemented by a simple Kramer Drive. The Slip Power thus harnessed can be pumped back to the supply bus bars through a step up Transformer.

This has resulted in energy savings of the order of 360 kW worth $300,000 per annum for a 5000 HP Wound Rotor Motor running at 90% full speed in the grinding Mills used in large Cement Plants [3]. Also, as the starting Torque developed by the Wound Rotor Motor is directly proportional to 'Slip", the Machine develops higher starting Torque as compared to a Squirrel Cage Motor at lower values of Current. Hence, a Drive system is designed and analysed with Wound Rotor Machine as Traction Motor in this paper.

2. SALIENT FEATURES OF WRIM

With the advancement in technology in construction of Electrical Machines, large WRIM are being manufactured of the rating of 18,000kW with a voltage range of 13,800V by leading manufacturers like ABB. These machines are manufactured to work at near unity power factor and at about 95% efficiency. The WRIMs have high starting torque, high inertia and low starting current with an enhanced feature of producing high torque over entire speed range. The machines come with automatic Brush lifting gears in which, the slip rings made of stainless steel having a smooth and non-grooved surface. After attaining the full rated speed, the Brush Lifting and Slip Ring short Circuiting Gear short circuits the Rotor and then lifts the brushes from Slip Rings. The other design is to have permanent contact brushes wherein the Slip Rings are manufactured from highly corrosion resistant Copper-Tin-Nickel alloy and helically grooved. The WRIM of ultra high ratings also come with self-ventilation with a fan mounted on the machine Shaft itself. Hence, adequate cooling is provided for safe and reliable operation [4]. The above mentioned features of WRIM are of advantage from the Locomotives point of view as the Traction Motors need to be robust to with stand vibrations, need to have better cooling for reliable and safe operation, need to have high starting Torque with low starting current and need to save on the reactive Power consumed as it needs to higher Power Bills.

The added feature of Slip-Power Recovery scheme in a WRIM would result in substantial savings in energy and hence reduce the cost of operation. The Slip-Power Recovered can be utilised to run the Locomotive Auxiliary units like the Blower Motors, for Lighting the cab and Machine units, charging of Battery used for raising the Pantograph etc. There is also a demand for Head-On-Generation of power for power the trailing Passenger Cars in Express trains like Rajdhani/Shatabdi/Durontos in order to bring down the dependence on Diesel Generators used in End-On-Generation [5].

3. SLIP-POWER RECOVERY SCHEME

The efficiency of an Asynchronous Motor is considerably reduced because of the presence of 'Slip'. The Slip Power gets wasted as heat in the Rotor of the Machine. This power can be harnessed in the case of a WRIM by means of either Static Kramer Drive or Static Scherbius Drive. The difference between the two is that a Scherbius Drive is a bi-directional Drive wherein power flow can be in either direction. If super-synchronous speed is required, power can be injected into the Rotor Circuit from the Bus-Bars. The Power thus harnessed can be used to Drive Auxiliary Loads of a Locomotive instead of being fed back to the Bus-Bars through Step-Up Transformer as this could add to the initial Cost. It has been found that the PWM Slip-Power Recovery does not create torque ripples in the Rotor of WRIM and hence, the efficiency of the Motor is not compromised [3]. In this paper a simple modified Kramer Drive is used to harness the Slip Power from the Rotor of the WRIM.

4. DESCRIPTION OF PROPOSED CIRCUIT

The proposed circuit with its detailed circuit Diagram is discussed in this section. For the purpose of simplicity the Circuit will be broken up into various parts like Supply system, Rectifier, Inverter, Auxiliary Converter etc. Both Rectifier and Inverter used in this paper use Insulated Gate Bi-polar Transistor (IGBT) as the switching device as the IGBTs have been found to have better thermal capabilities and also have been found to withstand sudden short-circuit conditions as compared to GTOs [6]. The other reasons for using IGBT as switching device are that soft-switched IGBTs have lower turn-on and turn-off losses, they require lower switching power and are capable of switching at high speeds [7]. The proposed circuit diagram is shown in Figure 1.

4.1. The Traction Supply System

The Traction Supply system consists of an AC Voltage source supplying 25kV AC at a frequency of 50Hz. This is to simulate the 25kV supply supplied from the Over-Head Equipment in the Indian Railway System. The supply is fed to a multi-winding Transformer with a primary rating of 25kV and secondary rating of 4500 V. The secondary of the Transformer is connected to Traction Rectifier.

4.2. Traction Rectifier System

The main rectifier is a parallel connected two 4 – Pulse Bridge units forming one of Traction Rectifier. They are two sets of such Traction Rectifier systems in this proposed circuit diagram, each individually fed from the secondary of the Traction Transformer. Eight IGBTs are used to form the Traction Rectifier Unit. The input to the Rectifier is 4500 V, 50 Hz AC supply fed from secondary of the Main Transformer. The output of the Rectifier is fed to the DC Link.

4.3. DC Link

The output of Traction Rectifier is connected to the DC Link. The DC Link consists of a Capacitor Bank of 815μF and 11.41 mF connected in parallel to filter out the Harmonics in the DC Voltage. A Diode is connected in the DC Link to ensure unidirectional current. A Braking Rheostat is connected in the DC Link through an IGBT. This acts as the Braking Chopper. The purpose of this Resistor is to introduce Dynamic Braking. If Braking is to be introduced, then the Circuit Breaker in the DC Link on the Inverter side can be opened by an external command and the Braking Resistor is inserted into the circuit by delivering Pulses from the Pulse Generator to the IGBT. The Pulse Generator is also controlled externally so that Braking can be done at anytime. If the Motors are to be accelerated again, the Pulses to the IGBT connected to the Braking Resistor are stopped and the Circuit Breaker on the Inverter side of DC Link can be closed. The output of the DC Link is connected to the Traction Inverter.

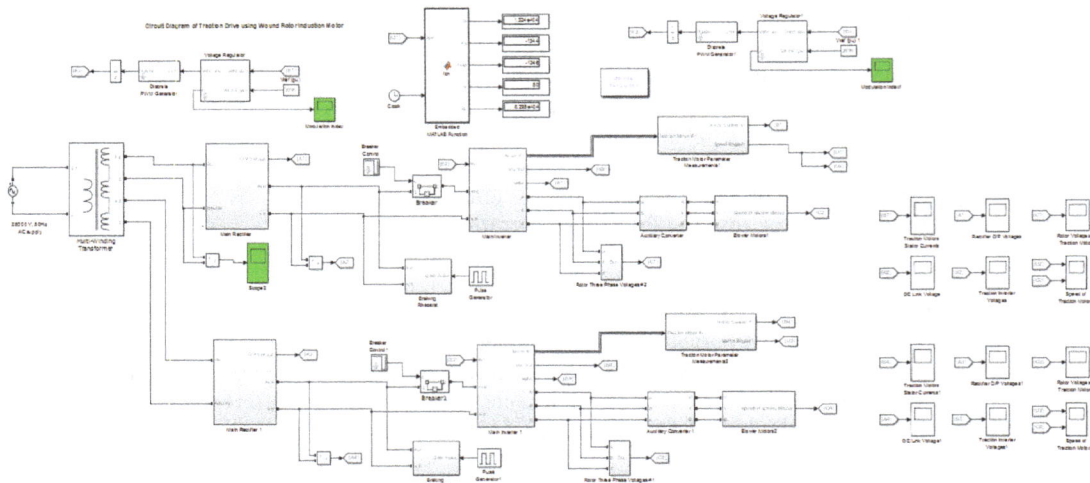

Figure 1. Proposed Circuit Diagram of the Traction Drive System using WRIM

4.4. Traction Inverter

A constant V/f Variable Voltage Variable Frequency control of Induction Motor will match the supply and demand torque by eliminating the use of a Flywheel [8]. Hence a three-phase Inverter is proposed in this paper.The Traction inverter is a 6-Pulse Bridge Inverter circuit which is capable of generating Sine waves displaced by a phase difference of 120°. The Pulses are delivered by means of a PWM generator. The output of the Traction Inverter is fed to the Traction Motors. The control of the Inverter is achieved by means of a simple constant V/f Technique where in the Three-Phase Voltages of the Inverter are measured and compared with the reference value of Voltage required and then the required frequency of Pulses are generated by the PWM generator so as to maintain a constant torque even at lower speed. The system is designed in such a way that, a Traction Inverter will supply Power to Two Traction Motors connected in parallel. This means that, for a 6-Axle Locomotive, there will be three Traction Inverters feeding the Motors. This will ensure 100% reliability in operation of the Locomotive.

4.5. The Traction Motor Circuit

An Asynchronous Motor with Wound Rotor is used as Traction Motor. The present circuit consists of a set of six Traction Motors one each for an Axle. The Motors are controlled from the Stator side for the simplicity of operation. The Motors are rated at 900kW, 3000 V, 50Hz with 2 Poles. The rated speed is 3000 rpm. The Rotors of the three Traction Motors are connected to one Auxiliary Converter, thus making two sets of Auxiliary Converters for the six Traction Motors. The speed is continuously measured and given as input to the Embedded MATLAB function which calculates the Power generated in the Air-Gap, the Slip-Power, the Torque developed and the Tractive Effort developed continuously. The required frequency according to the new speed required is also calculated in the Embedded MATLAB function.

4.6. The Auxiliary Converter

The Auxiliary Converter is basically Static Kramer Drive. It consists of a Three-Phase Rectifier with Diodes as switch. The Inverter is connected via a DC Link. The Inverter is an IGBT based Inverter. Hence, the Auxiliary Converter is uni-directional Converter capable of transferring Power from the Rotor Circuit of the Traction Motors to the Load. The Inverter is fired using a simple PWM Generator. The Auxiliary converter in the proposed design can drive a Load of six 15kW, 400 V, 50 Hz, 4-pole Squirrel Cage Induction Motors. These kinds of Three-Phase Squirrel Cage Machines are normally used as Blower Motors for cooling the Traction Motors. The circuit diagram shown in Figure 1 is simulated using MATLAB Simulink. The simulation results are discussed in the next section in detail.

5. SIMULATION AND RESULTS

The circuit that has been proposed in the Figure 1 was simulated using MATLAB Simulink software. The simulation was carried out for 10 seconds of Simulation Time to study the results in detail. The Tractions Motors were accelerated for a time period of 5 seconds. They reached the steady state speed in about 0.2 seconds. The Circuit Breaker on the Inverter side of the DC Link was opened and the Braking Resistor was introduced into the circuit by delivering pulses to the IGBT connected to the Braking Resistor at time of 5 seconds. The speed of the Traction Motors reduced to zero and went into super-synchronous speed region. Again at a time of 7 seconds, the Circuit Breaker of the DC Link on the Inverter side was closed and the pulsations to the IGBT connected to the Braking Resistor were ceased. This resulted in the Traction Motors accelerating again to the required speed.

5.1. The Traction Rectifier Output

The output waveform of the Traction Rectifier is shown in Figure 2. The Waveforms were observed to be ripple free and with fewer harmonic. The amplitude of the output Voltage was 5000 V. The output Voltage was a pulsating DC waveform. The waveform obtained has been zoomed for better view in the Figure.

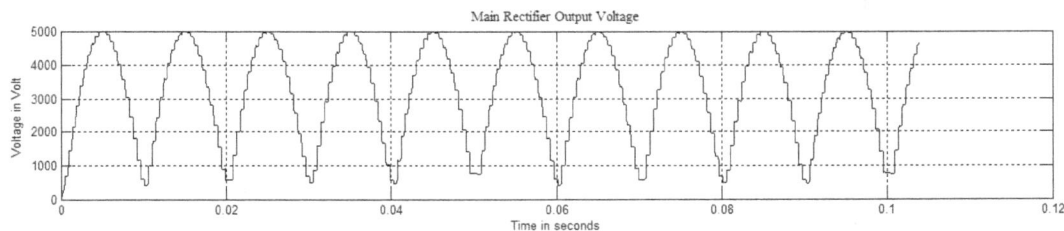

Figure 2. Traction Rectifier Output Voltage Waveform

5.2. DC Link Output

The Capacitors of values 815μF and 11.42 mF were connected in parallel to form the Capacitor bank to filter out the Harmonics and also work as a Voltage Booster. The Diode was connected to ensure uni-directional Power flow. The output Voltage waveform is shown in Figure 2. The Voltage waveform was observed to be pure straight line DC of the amplitude of 5800 Volts. The variation of amplitude of DC Link Voltage can be observed in the Graph at t = 5 seconds.

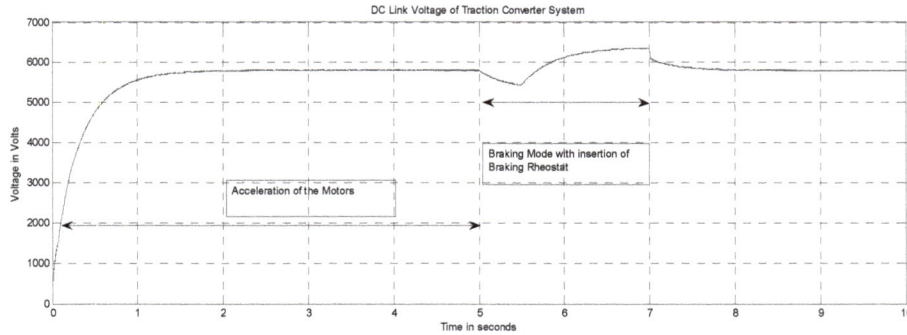

Figure 3. DC Link Output Voltage Waveform

5.3. Traction Inverter and Motors Outputs

The output of DC Link is connected to the Traction Inverter. The Traction Inverter is pulsed by the Discrete PWM Generator based on simple constant V/f principle [4]. The no-load voltage to rated frequency ratio is calculated in the Embedded MATLAB function. The frequency of operation of the Inverter is changed to change the speed of the Traction Motors. The Speed of the Traction Motor is fed to the Embedded MATLAB function. The new speed required is given as a command during run-time at a pre-defined time in the Embedded MATLAB function. The new frequency of firing corresponding to the new required speed is calculated. The Voltage boost required for the new frequency is also computed from the V/f ratio. These inputs are fed to the PI Controller to regulate the Voltage Regulator block. The new Frequency required and the corresponding Voltage required is compared and firing pulses are given to the Traction Inverter. The Voltage level is varied so as to maintain the Torque constant. The Waveforms of Three-Phase Inverter output Voltages are shown in Figure 4.

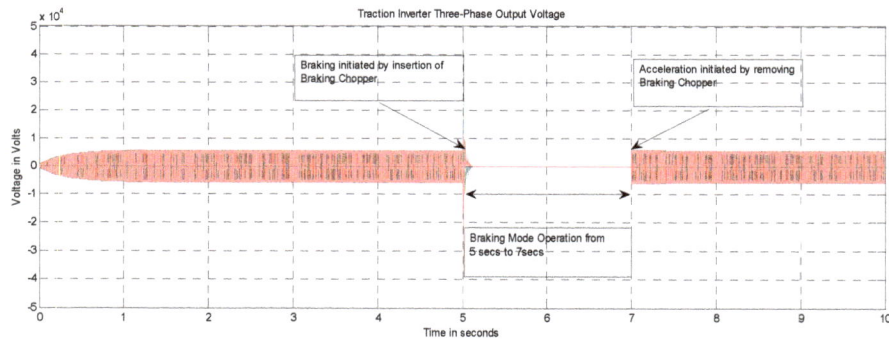

Figure 4. Traction Inverter Three-Phase Output Voltage Waveforms

The Phase-Phase Voltage was 4500 V and the current was 100 Amp continuous. The Braking Chopper was pulsed at t=5 seconds with the opening of the Circuit Breaker. The Inverter Voltages and Currents dropped to zero and the current circulated in the DC Link through the Braking Chopper. At t=7 seconds, the Circuit Breaker was closed and the pulses to the Braking Chopper were stopped. This resulted in Traction Motors accelerating again.

The Traction Motors achieved steady state speed at t=0.2 seconds. After reaching the steady state, the Traction Motors ran at near rated speed of 3000 rpm. The speed observed for the Traction Motors in continuous mode of operation was 2900 rpm with minor oscillations. With the introduction of the Braking Chopper at t=5 seconds, the speed dropped to zero. The Motors accelerated again to near rated speed after the Braking Chopper was removed from the circuit at t=7 seconds. The Speed curve of the Traction Motors is shown in Figure 5.

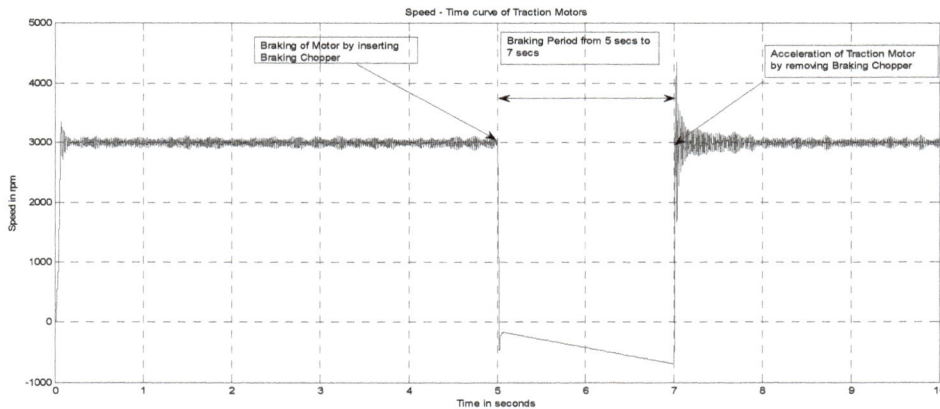

Figure 5. Speed of the Traction Motors

5.4. The Auxiliary Converter Output

The Three-Phase Rotor output Voltages of the Traction Motors were fed to the Three-Phase Auxiliary Rectifier. The Rectifier was a Diode Rectifier and hence there was no bi-directional flow of current. The Three-Phase Rotor Voltages of the Traction Motors are shown in Figure 6. The Phase – Phase Voltage of 300V was observed. The Capacitor Bank in the DC Link of the Auxiliary Converter of the capacity 815μF filtered out the harmonics and also boosted the DC Link Voltage to 400V. The Three - Phase IGBT Auxiliary Inverter developed a Voltage of 450V ph-ph and this was fed to the Blower Motors.

Six Squirrel Cage Motors were connected to the circuit in two sets of three Motors each. The Blower Motors could achieve the rated speed of 1500rpm in about 4.5 seconds during which the Traction Motors accelerated to 3000rpm. The continuous speed observed for the Blower Motors was 1490rpm. As expected from the theoretical calculations, the Slip Power recovered from the Traction Motors was able to drive the Blower Motors with the Blower Motors achieving the desirable rated speed.

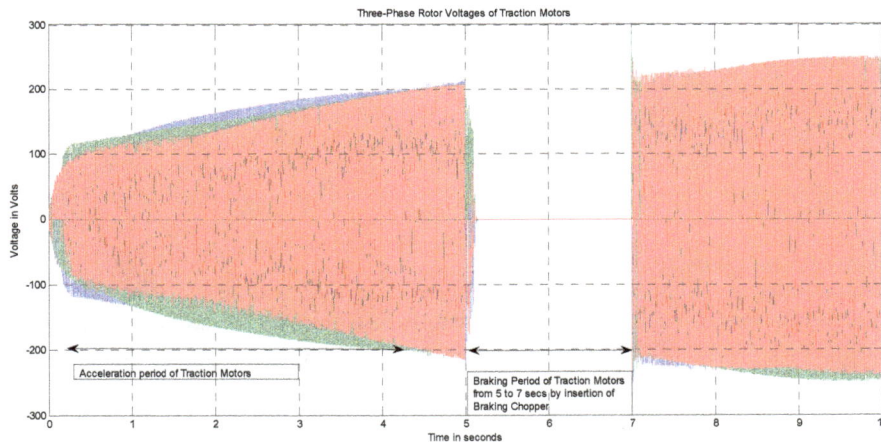

Figure 6. Three – Phase Rotor Voltages of the Traction Motors

The Speed Curve observed for the Blower Motors when Braking Chopper was introduced in the Traction Converter system is shown in Figure 7. It was observed that the Blower Motors took longer time duration to reach the steady state speed and rated speed as compared to the time duration taken when there was no Braking Chopper in the circuit. The Blower Motors took 15 seconds to reach the near rated speed of 1490rpm with Braking Chopper.

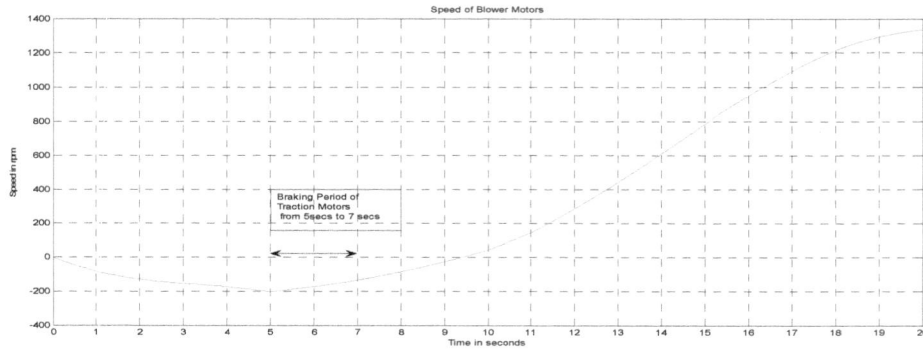

Figure 7. Speed of the Blower Motors in rpm with Braking of Traction Motors

6. EQUATIONS AND CALCULATIONS

The Equations and Calculations related to the proposed Traction Drive systems are presented in this section in brief.

6.1. Calculation of Tractive Effort Required

The various Tractive Efforts required by a Locomotive are:
Tractive Effort for Acceleration (F_a):

$$F_a = 277.8 \, W_e \, \alpha \, Newtons \tag{1}$$

Tractive Effort to overcome Gravitational Pull (F_g):

$$F_g = 9.81 \, W.G \, Newtons \tag{2}$$

Tractive Effort required to overcome Train Resistance for a Locomotive (F_r):

$$F_r = 9.81(\, 0.65 \, W_l + \, 13 \, n + 0.01 \, W_l \, v + 0.52 \, v^2 \,)Newtons \tag{3}$$

Tractive Effort required to overcome Curve Resistance (F_c):

$$F_c = 9.81 \, W \left(\frac{700}{R}\right) \, Newtons \tag{4}$$

Total Tractive Effort = $F_t = F_a + F_g + F_c + F_r$ \hfill (5)

6.2. Assumptions for Calculations

It is assumed that the Locomotive starts on a plane surface without Gradient and Curvature hence, the Tractive Effort required would be only Tractive Effort for Acceleration. Let us assume that the Locomotive has to accelerate a trailing Load of 1500 tonne to 120 Kmph in 400 seconds.

(a) Calculation of Tractive Effort:

Acceleration, α in Kmphps will be given as, $\alpha = \frac{120}{400} = 0.3 \, Kmphps$

Weight of the Locomotive = W_l = 123 tonnes
Weight of the Trailing Load = W_t = 1500 tonnes
Total Weight = W = (W_l + W_t) = 1623 tonnes
Effective weight of Locomotive and Trailing Load = W_e = 1785.3 tonnes
Tractive Effort required for Acceleration = F_a = 277.8 × 1785.3 × 0.3 = 149 $kNewton$

(b) Calculation of Power, Torque developed:
Outputs from the Embedded MATLAB function,
Torque developed by one Traction Motor = 13240 Nm
Air-Gap Power of one Traction Motor = 840 kW
Slip = 0.03

Slip – Power of one Traction Motor = $0.03 \times 840 \ kW = 25.2 \ kW$
Tractive Effort developed by one Traction Motor = 30 kN
Rotor Voltage developed at the Rotor of Traction Motor = 300 V
Output Voltage of Auxiliary Converter = 450 V
Power consumed by each Blower Motor = 15 kW

If the Locomotive works for 15 hours a day and maintains the rated speed for atleast 10 hours during running, the Slip-Power energy recovered will be $S_p = 10 \times 15 \ kW = 150 \ kWH$ per Traction Motor.

For a period of 300 days of running of Locomotive in a Calendar year, the total Slip-Power energy recovered will be $S_{ptotal} = 6 \times 150 \ kWH \times 300 \ days = 270000 \ kWH$.

One Unit of energy is sold at an average of Rs. 8 in India and hence the amount of money saved by Slip-Power Recovery Scheme would be $= 8 \times 270000 = Rs. 2160000/-$ per annum per Locomotive.

For a Fleet of 2000 Electric Locomotives, the amount of money saved would be $= 2000 \times 270000 = Rs. 54,00,00,000/-$ per annum

Table 1. Traction and Blower Motors Parameters:

S. No	Parameter	Traction Motors	Blower Motors
1.	Rated Voltage	3000 V	400 V
2.	Rated Power	900 kW	15 kW
3.	Operating Frequency	50 Hz	50 Hz
4.	Efficiency	95 %	95 %
5.	Power factor	0.85 lag	0.85 lag
6.	No. Of poles	2 poles	4 poles

7. CONCLUSION

From the Simulation studies carried out with MATLAB for the above proposed circuit, it can be concluded that the use of a Wound Rotor Induction Motor would result in enormous amount of savings in energy. As the Slip-Power recovered from the Traction Motors will be able to drive the Blower Motors, the use of power drawn from the Over-Head 25kV supply through Auxiliary winding of Main Transformer for driving the Blower Motors can be reduced resulting in reduction of cost of buying Power from the Distribution Companies. Though the initial cost of the Wound Rotor Induction Motor and the required enhanced circuitry may be high, it can be recovered in few years time during the life of a Locomotive, which is about 40 years. This scheme can be further studied and implemented as the future technology for Electric Traction Systems in India, where the Railways are aiming at cost cutting initiatives to enhance the profitability of running the whole network.

REFERENCES

[1] Matthew P Magill, Phillip T Krein. Examination of Design Strategies for Inverter-Driven Induction Machines. *Power and Energy Conference Illinois (PECI) 2012 IEEE*. 2012; 1-6.
[2] H Partab. Modern Electric Traction. *Publisher: Dhanpat Rai and Sons*, India. 2012.
[3] Paul Blaiklock, William Horvath. Saving Energy. *TMEIC GE, USA – Motor Technology*. 2009.
[4] ABB Motors, Generators. Brochure on Slip Ring Motors for heavy-duty and critical Applications. 2011.
[5] J Upadhyaya, SN Mahendra. Electric Traction. *Allied Publishers India*. 2000. ISBN: 10:8177640054.
[6] Hansruedi Zeller. High Power Components from the State of the Art to Future Trends. *Power Conversion (PCIM 1998)*. 1998; 1-10.
[7] J Arrillaga, YH Liu, NR Watson and NJ Murray. *Self-Commutating Converters for High Power Applications*. John Wiley & Sons, Ltd. 2009 ISBN: 978-0-470-74682-0
[8] SS Chirmurkar, MV Palandurkar, SG Tarnekar. Torque Control of Induction motor using V/f Method. *International Journal of Advances in Engineering Sciences*. 2011; 1(1).
[9] Toby J Nicholson. *DC and AC Traction Motors*. IET Professional Development Course in Traction Systems. 2008; 34-44.
[10] Rupesh Kumar. Course on Three Phase Technology in TRS Applications. *IRIEEN, Nasik, India*. 2010.
[11] C Bharatiraja, S Raghu, Prakash Rao, KRS Palanisami. Comparative Analysis of Different PWM Techniques to reduce Common mode Voltage in Neutral Point Clamped Inverter for Variable Speed Induction Drives. *International Journal of Power Electronics and Drive System (IJPEDS)*. 3(1): 2013: 105~116.
[12] R Rajendran, N Devarajan. A Comparative Performance Analysis of Torque Control Schemes for Induction Motor Drives. *International Journal of Power Electronics and Drive System (IJPEDS)*. 2012; 2(2): 177~191.
[13] MATLAB Simulink 2014a *MathWorks Inc. USA*. 2014.

22

The Switched Reluctance Electric Machine with Constructive Asymetry

A. Petrushin, M. Tchavychalov, E. Miroshnichenko
Chair "Electric Rolling Stock", Rostov State Transport University, Rostov-on-Don, Russia

ABSTRACT

Keyword:

Magnetic system
Noise
Switched reluctance machine
Unbalanced unilateral attraction
Vibration

Results of researches of forces of a unilateral attraction of a rotor to a stator of switched reluctance electric machines taking into account unevenness of an air gap are given. It is offered for configurations of magnetic systems with weak magnetic communication between the coils making a phase to carry out a supply of coils in parallel or independently from each other for reduction of not compensated forces of a unilateral attraction of a rotor to a stator.

Corresponding Author:

M. Tchavychalov,
Chair "Electric Rolling Stock",
Rostov State Transport University,
2 Narodnogo Opolcheniya sq., Rostov-on-Don, 344038, Russia.
Email: chavychalov-maxim@yandex.ru

1. INTRODUCTION

Theory of switched reluctance machines (SRM) was in good progress. As a result SRM technical and economic indexes are on a level of the best traditional electric machines with circular air gap magnetic field. However there are unsolved problems. One of the most difficult to solve is vibroacoustic indexes development.

Taking into account the lost in value of high quality power electronic it is possible now to set and to solve new problems of SRM electromechanical characteristics development. The improvement of vibroacoustic indexes is among of them.

Analytical review of SRM noise and vibration reduction implies that problem is intricate and do not have simple solutions. Umbrella approach is necessary to find out the factors acting on the SRM vibration and noise. One of those factors is force of unilateral attraction rotor to stator. It is variable and takes place practically always. The main cause of unilateral attraction is margin tolerance on parts of SRM. The force of unilateral attraction is more in small air gap SRMs because of greater dissymmetry of magnetic system [1].

2. RESEARCH METHOD

SRM magnetic systems are distinct in number of phases, tooths and coils. The most of SRM has weak magnetic coupling between phases and the magnetic flux is common for all tooths of one phase (Figure 1).

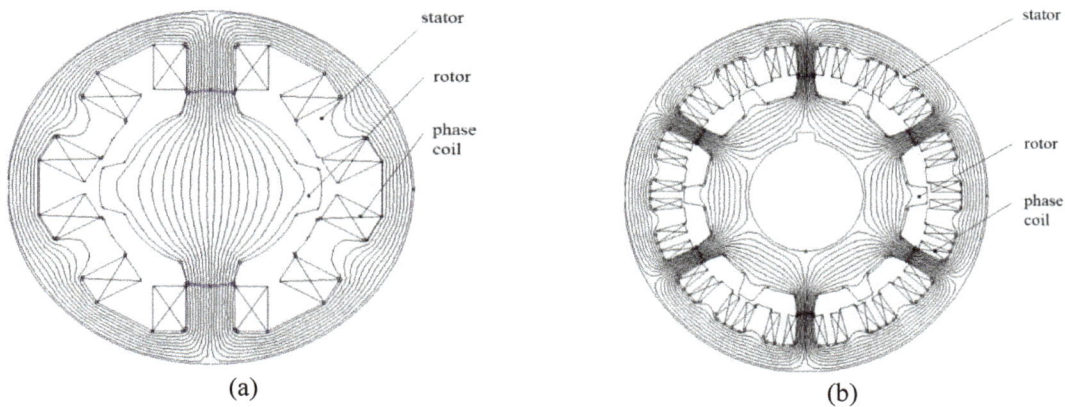

Figure 1. Distribution of magnetic field lines of SRM; a) 6/4, b) 18/12

SRM rotor is always shifted in no matter which side within tolerance. When the air gap is rather small even small shift of the rotor is the cause of occurrence of the forces of unilateral attraction stator to rotor. It is difficult to cancel out those forces by current regulation in separate coils because of common flux in traditional SRM configurations. However, it is possible to do in "short-flux" SRM where separate phase coils have separate flux [2], [3]. It will reduce the bearing assembly load and will increase the margin of safety.

Let us consider magnetic system on Figure 2 [2]. Phase coils are set at an 120 degrees angle and produce oncoming fluxes.

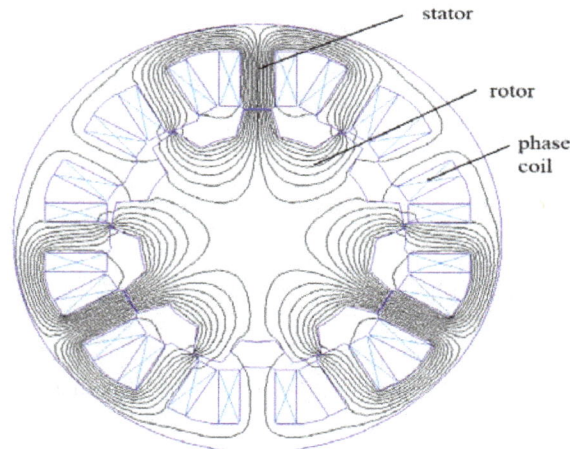

Figure 2. Distribution of magnetic field lines of SRM 12/9

It is rational to connect phase coils in parallel with power supply in SRM configurations like shown at Figure 2. In that case constructive dissymmetry will be compensated by currents in the phase coils. Smaller air gap means less coil inductance. As a result coil current will grow faster in the coil with the smaller air gap. Different current in phase coils will produce the radial force counterbalancing the rotor.

The calculations were carried out for magnetic system with connection phase coils in parallel and with connection in series. To create the constructive unbalance the rotor was vertically moved. For example moving the rotor in 0.05 mm (25% of nominal air gap) in align position the inductance of coil A is 2.332 mH and the inductance of coils B and C is 1.921 mH. The difference in inductance is 17%. Numeric values of coil inductance were computer calculated by finite elementary method.

Mathematical model of SRM phase with coils connected in series is as follows [4]-[6]:

$$
\begin{cases}
\dfrac{di}{dt} = \dfrac{1}{L(i,\theta)}\left(v - i \cdot R - \omega \cdot \dfrac{\partial \psi(i,\theta)}{\partial \theta}\right), \\[3mm]
\dfrac{d\omega}{dt} = \dfrac{1}{J}\left(T - B \cdot \omega - T_L\right), \\[3mm]
\dfrac{d\theta}{dt} = \omega,
\end{cases}
\tag{1}
$$

Where: i – phase current, A;

L(i,θ) – phase inductance, H;

θ – rotor position, rad;

v – voltage, V;

R – phase resistance, Ohm;

ω – rotation frequency, rad/s;

ψ(i,θ) – flux linkage, Wb;

J – equivalent moment of inertia, kg·m2;

B – friction ratio;

T – electromagnetic torque, Nm;

T_L – load torque, Nm;

When the phase coils are connected in parallel there is no magnetic coupling between phase coils. It is possible to calculate phase coil currents independently. Phase voltage is applied to all phase coils at the same time. Consequently for coil currents it can be written:

$$
\begin{cases}
\dfrac{di_a}{dt} = \dfrac{1}{L_a(i,\theta)}\left(u - i_a \cdot R_{pc} - \omega \cdot \dfrac{\partial \psi_a(i,\theta)}{d\theta}\right), \\[3mm]
\dfrac{di_b}{dt} = \dfrac{1}{L_b(i,\theta)}\left(u - i_b \cdot R_{pc} - \omega \cdot \dfrac{\partial \psi_b(i,\theta)}{d\theta}\right), \\[3mm]
\dfrac{di_c}{dt} = \dfrac{1}{L_c(i,\theta)}\left(u - i_c \cdot R_{pc} - \omega \cdot \dfrac{\partial \psi_c(i,\theta)}{d\theta}\right), \\[3mm]
\dfrac{d\omega}{dt} = \dfrac{1}{J}\left(T_a + T_b + T_c - B \cdot \omega - T_L\right), \\[3mm]
\dfrac{d\theta}{dt} = \omega,
\end{cases}
\tag{2}
$$

Where: Rpc – phase coil resistance, Ohm.

3.   RESULTS AND ANALYSIS

To compute SRM parameters by Equation (1) and (2) the accepted assumption is to neglect the mutual phase coupling. For magnetic system under discussion (Figure 2) magnetic phase coupling is usually taken into account. However to receive qualitative result of comparison of series and parallel phase coils connection that assumption can be accepted because only the difference of coil currents is estimated.

Calculations were carried out in SIMULINK/MATLAB [7]. SRM phase modes for series and parallel connection of phase coils are shown in Figure 3 and 4 as follows.

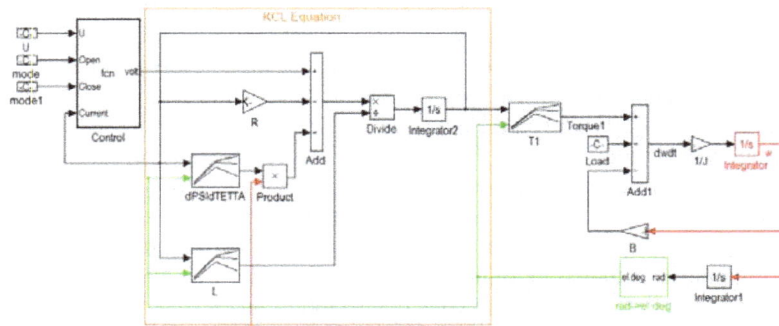

Figure 3. SIMULINK model of SRM phase with series-connected coils

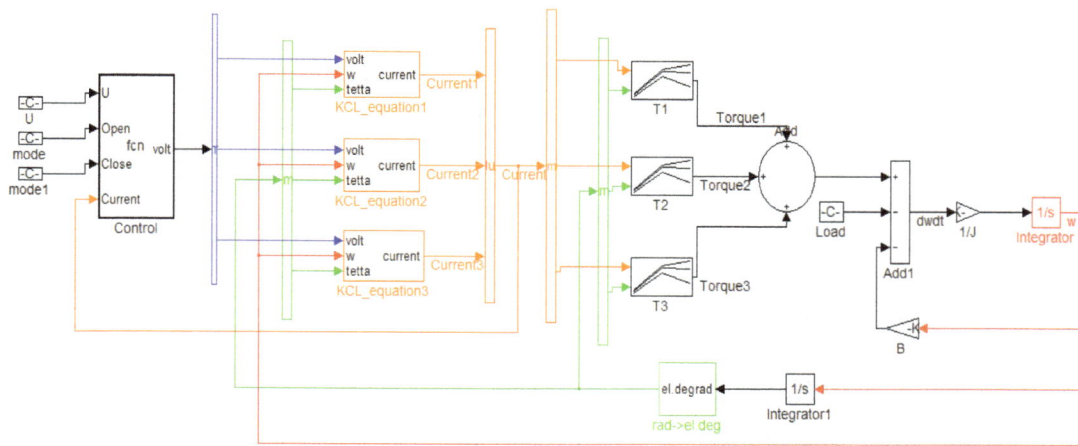

Figure 4. SIMULINK model of SRM phase with parallel-connected coils

Initial data for computations in the form of table $\psi(i,\theta)$ were received for SRM rotor unilateral shift to 25%, 50% and 75% of designed air gap (0,05; 0,1 и 0,15 mm). The simulation was carried out for nominal load SRM operation on a rotation frequency of 100 rad/s. DC link voltage for series phase coils connection was 60 V, for parallel connection – 20 V. As a result the dependencies $i(\theta)$ were received (Figure 5).

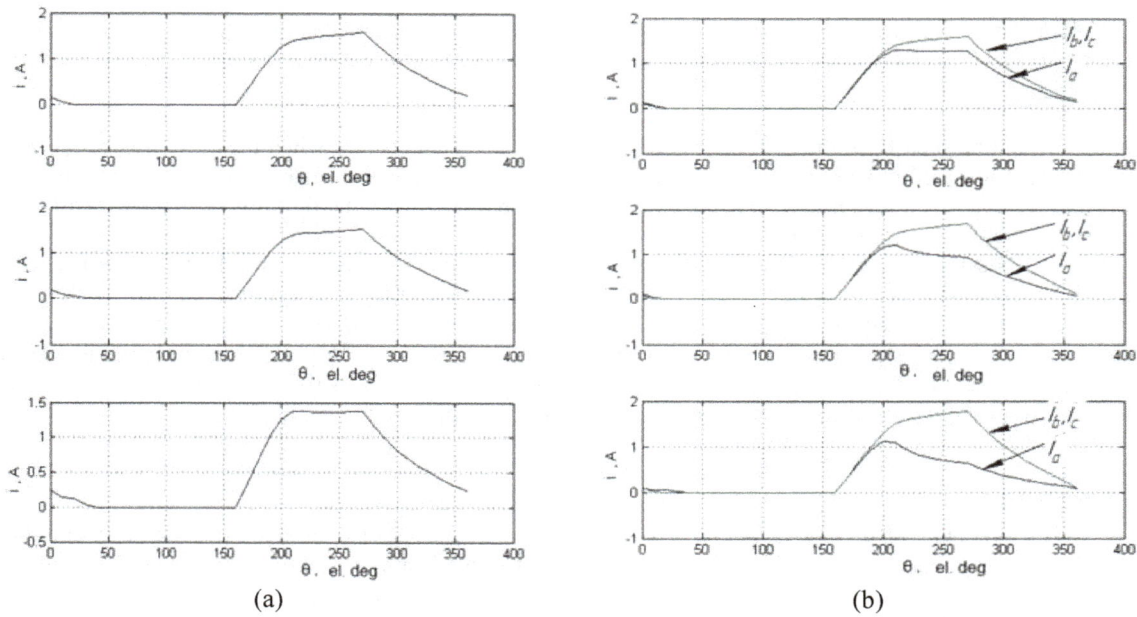

(a)   (b)

Figure 5. Results of SRM modeling (a – series-connected phase coils, b – parallel-connected phase coils) under rotor shift 0.05, 0.1 and 0.15 mm

Subsequent to the results of SRM phase modeling forces of unilateral rotor attraction were calculated by finite elementary method (Figure 6).

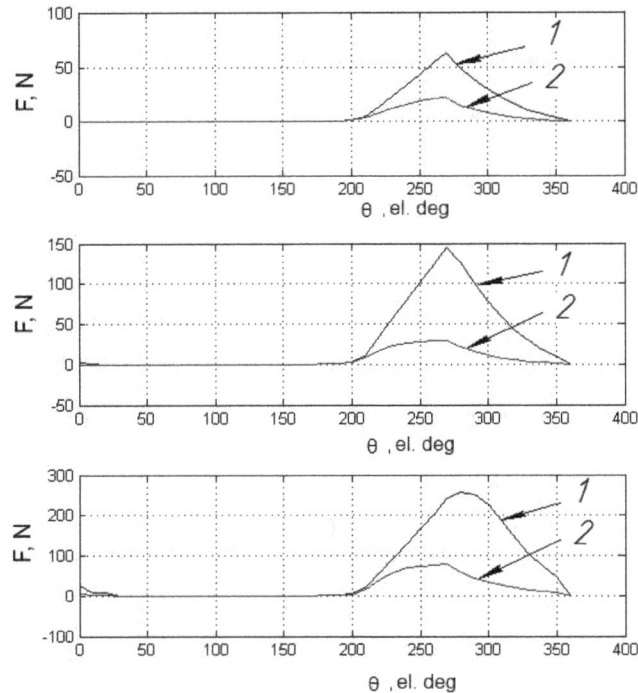

Figure 6. The force of unilateral rotor to stator attraction (1 – series-connected phase coils, 2 – parallel-connected phase coils) under rotor shift 0.05, 0.1 and 0.15 mm

As can be seen from the Figure 6 with series phase coils connection when the current is equal in all phase coils the force of unilateral rotor attraction is rather more than in case of parallel phase coils connection when coil current depend from air gap distance. Connecting phase coils in parallel it is possible to reduce the force of unilateral rotor attraction in 3 times at the average.

It is possible to reduce the force of unilateral rotor attraction in SRM with relatively low magnetic coupling between phase coils by operating in artificial magnetomoving force dissymmetry mode. In this case every phase coil is supplied by singular semiconductor switch. It is required to increase current in coil where the air gap distance is larger and to decrease current in coil with smaller air gap. It is also possible in closed-loop control system using the sensor of radial displacement to implement active magnetic levitation of the rotating rotor and to unload the bearings. Thus the vibration and noise will be reduced.

4. CONCLUSION

The specialty of the SRM magnetic systems wherein the phase coils have no magnetic coupling render possible to reduce the force of unilateral attraction rotor to stator connecting the phase coils in parallel. In this case the negative feedback between the air gap distance and unbalanced force of unilateral rotor attraction will be valid. In the case of supplying phase coil by singular semiconductor switch it is possible to compensate the forces of unilateral attraction almost completely in the same way that in active magnetic bearing. Creating conditions for bearing unload is especially relevant for high-duty electric machines operating in extreme duty cycles. Proposed solutions will make possible to reduce the SRM vibroactivity and noise as consequence.

ACKNOWLEDGEMENTS

The work is supported by The Ministry of education and science of Russia. Agreement № 14.604.21.0040, identifier RFMEFI60414X0040.

REFERENCES

[1] Petrushin AD, Iljasova EE, Investigation of the effect of uneven air gap on the value of unilateral attraction force of the rotor to the stator switched reluctance electric machine. *Newspaper of VElNII*, 2011; 2: 84-93.

[2] Petrushin AD, Grebennikov NV. Reactive switched electric machine with rotational symmetry. *Patent RF №2450410,* 2012.

[3] Miller TJE, Hendershot JR. Design of Brushless Permanent-Magnet Motors, *Magna Physics Publishing and Glarendon Press.* OXFORD. 1994; 512.

[4] Krishnan R. Switched reluctance motor drives: modeling, simulation, analysis, design, and applications. *Magna Physics Publishing.* 2001; 416.

[5] Ghousia SF. Impact analysis of dwell angles on current shape and torque in switched reluctance motors. *International journal of power electronics and drive systems.* 2012; 2(2): 160-169.

[6] Srinivas P, Prasad VN. Direct Instantaneous torque control of 4 phase 8/6 switched reluctance motor. *International journal of power electronics and drive systems.* 2011; 1(2): 121-168.

[7] Wadnerkar VS. Performance analysis of switched reluctance motor; design, modeling and simulation of 8/6 switched reluctance motor. *Journal of Theoretical and Applied Information Technology.* 2008; 11: 1118-1124.

23

Analysis of Variable Speed PFC Chopper FED BLDC Motor Drive

A. Jeya Selvan Renius, K. Vinoth Kumar
Department of Electrical and Electronics Engineering, School of Electrical Sciences,
Karunya Institute of Technology & Sciences University, Coimbatore – 641114, Tamilnadu, India

ABSTRACT

Keyword:

Bridgeless
Common-mode noise
DC-DC chopper
Power factor correction
Power quality

This paper provides the detailed analysis of the DC-DC chopper fed Brushless DC motor drive used for low-power applications. The various methods used to improve the power quality at the ac mains with lesser number of components are discussed. The most effective method of power quality improvement is also simulated using MATLAB Simulink. Improved method of speed control by controlling the dc link voltage of Voltage Source Inverter is also discussed with reduced switching losses. The continuous and discontinuous modes of operation of the converters are also discussed based on the improvement in power quality. The performance of the most effective solution is simulated in MATLAB Simulink environment and the obtained results are presented.

Corresponding Author:

A. Jeya Selvan Renius,
Department of Electrical and Electronics Engineering,
School of Electrical Sciences, Karunya University,
Coimbatore – 641114, Tamilnadu, India.
Email: renius28@gmail.com

1. INTRODUCTION

Low power motor drives such as fans, water pumps, blowers, mixers, HVAC transmission, motion control etc. use BLDC motor for their efficient operation. Since BLDC offers high efficiency, low electromagnetic interference, low maintenance and high flux density per unit volume, we use BLDC for low power applications. BLDC motors are very popular in a wide variety of applications. Compared with a DC motor, the BLDC motor uses an electric commutator rather than a mechanical commutator, so it is more reliable than the DC motor. In a BLDC motor, rotor magnets generate the rotor's magnetic flux, so BLDC motors achieve higher efficiency. Therefore, BLDC motors may be used in high-end white goods (refrigerators, washing machines, dishwashers, etc.), high-end pumps, and fans and in other appliances which require high reliability and efficiency.

In this respect, the BLDC motor is equivalent to a reversed DC commutator motor, in which the magnet rotates while the conductors remain stationary. In the DC commutator motor, the current polarity is altered by the commutator and brushes.

However, in the brushless DC motor, polarity reversal is performed by power transistors switching in synchronization with the rotor position. Therefore, BLDC motors often incorporate either internal or external position sensors to sense the actual rotor position, or the position can be detected without sensors.

The choice of mode of operation of a PFC converter is a critical issue because it directly affects the cost and rating of the components used in the PFC converter. The continuous conduction mode (CCM) and discontinuous conduction mode (DCM) are the two modes of operation in which a PFC converter is designed to operate. In CCM, the current in the inductor or the voltage across the intermediate capacitor remains continuous, but it requires the sensing of two voltages (dc link voltage and supply voltage) and input side current for PFC operation, which is not cost-effective. On the other hand, DCM requires a single voltage sensor for dc link voltage control, and inherent PFC is achieved at the ac mains, but at the cost of higher stresses on the PFC converter switch; hence, DCM is preferred for low-power applications.

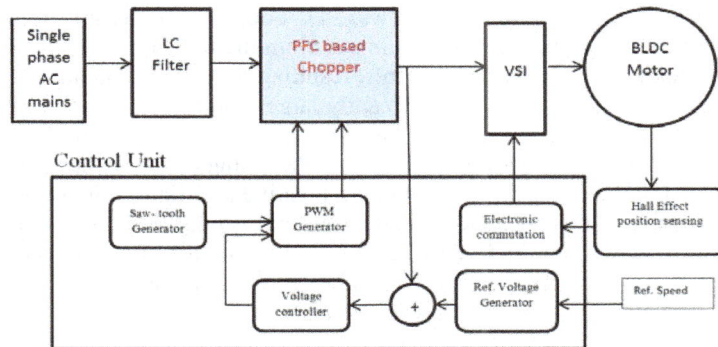

Figure 1. Block diagram of PFC chopper-fed BLDC motor drive

BLDC with diode bridge rectifier with a high value DC link capacitor has a THD (Total Harmonic Distortion) of 65% and power factor as low as 0.8. So the power factor is corrected using the PFC converters. Both continuous and discontinuous modes of the converters are discussed and the discontinuous mode of conduction is best suited for the low power applications. Since the discontinuous conduction requires only a single voltage sensor for DC link voltage control. But conventional PFC uses more number of components that increases the cost of the control circuit. Also the conventional PFC used PWM-VSI for speed control with constant DC link voltage which produces higher switching losses.

Thus the analysis is made for different methods that improve the power quality at the ac mains. For further improvement in efficiency, bridgeless (BL) converters are used which allow the elimination of DBR in the front end. A buck–boost converter configuration is best suited among various BL converter topologies for applications requiring a wide range of dc link voltage control (i.e., bucking and boosting mode). These can provide the voltage buck or voltage boost which limits the operating range of dc link voltage control. A new family of BL SEPIC and Cuk converters has been reported but requires a large number of components and has losses associated with it.

This paper presents a detailed analysis of chopper-fed BLDC motor drive with variable dc link voltage of VSI for improved power quality at ac mains with reduced components.

2. EXISTING TOPOLOGY

The conventional PFC uses Pulse Width Modulated Voltage Source Inverter (PWM-VSI) for speed control with constant DC link voltage. This causes higher switching losses. The switching losses in this conventional approach increases as a square function of switching frequency.

T. Gopalarathnam and H.A. Toliyat [1] in 2003 proposed a Single Ended Primary Inductance Converter (SEPIC) based BLDC which also has higher losses in the VSI due to conventional PWM switching and large number of current and voltage sensors are used that additionally adds to the cost of the converter.

S. Singh and B. Singh [2] in 2011 proposed a paper about Buck-Boost converter based on constant DC link voltage and also use PWM-VSI for speed control which again increases the switching losses.

S. Singh and B. Singh [3] again in 2012 proposed a cuk converter fed BLDC motor with a variable DC link voltage that reduces the switching losses since it uses only the fundamental switching frequency. Speed control is performed by controlling the voltage at the DC bus of VSI. In this paper, Continuous Conduction Mode (CCM) is used. But the major disadvantage is that it requires three sensors. So it is not encouraged for low-cost and low-power rating applications.

Since only the bridge converters are used in all the above used topologies, it also contributed for the switching losses. Thus the bridgeless topologies are preferred. The different bridgeless topologies are analysed based on the power quality of the ac mains.

3. BRIDGELESS CONVERTER TOPOLOGIES

The bridgeless converters eliminate the use of diode rectifiers. The diode rectifiers cause more switching stresses. This is not good for the proper functioning of the converter.

3.1. Boost Converters

Y. Jang and M.M. Jovanovic [4] in the year 2011 proposed a concept based on boost converter fed BLDC motor drive. The basic topology of the bridgeless PFC boost rectifier is shown in Figure 2. Compared to conventional PFC boost rectifier one diode is eliminated from the line-current path, so that the line current simultaneously flows through only two semiconductors, resulting in reduced conduction losses. However, the bridgeless PFC boost rectifier in Figure 2 has significantly larger common-mode noise than the conventional PFC boost rectifier. In fact, in the conventional PFC boost rectifier, the output ground is always connected to the ac source through the full-bridge rectifier whereas, in the bridgeless PFC boost rectifier in Fig. 2, the output ground is connected to the ac source only during a positive half-line cycle, through the body diode of switch, while during a negative half-line cycle the output ground is pulsating relative to the ac source with a high frequency (HF) and with an amplitude equal to the output voltage. This HF pulsating voltage source charges and discharges the equivalent parasitic capacitance between the output ground and the ac line ground, resulting in a significantly increased common-mode noise.

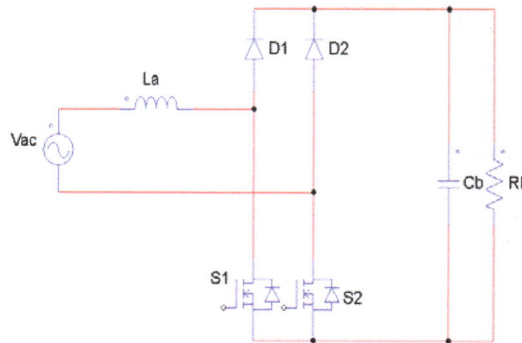

Figure 2. Bridgeless PFC Boost converter

The bridgeless boost converter provides only voltage boost which limits the operating range of DC link voltage control. Thus we move for another topology.

3.2. CUK Converters

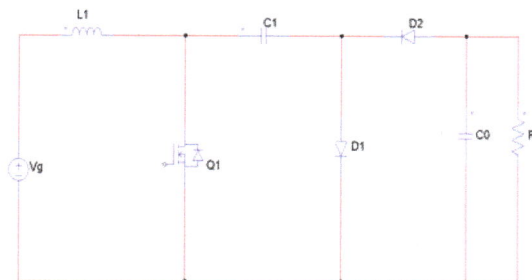

Figure 3. Modified Cuk converter with Negative output polarity

L. Huber, Y. Jang and M.M. Jovanovic [5] in the year 2008 proposed a paper based on cuk converter based BLDC. In this section, the topology derivation of the proposed converter is presented. Figure 3 shows a modified Cuk converter also known as a "Self-lift Cuk" converter. Referring to Figure 3, the converter can be manipulated to produce a positive output voltage from a negative input voltage. Similarly, for a converter it is possible to produce a negative output voltage from a negative input voltage. Note that the converters have similar output characteristics and they are identical except for their input voltage polarity and switch drain-to source connection. Therefore, it is possible to combine the two converters into a single bridgeless ac-dc PFC converter containing a bi-directional switch and an alternating input voltage source. Likewise, the converter can be combined into a single bridgeless ac-dc PFC converter which offers an inverted output voltage polarity. Unlike the conventional bridgeless PFC converters, all components in the proposed converter are fully utilized as there are no idle components during both the positive and negative ac-line cycle. Also, no additional diodes or capacitors are added to the topology to filter out common mode noise since the output is not floating.

This converter also has a serious disadvantage of switching losses. So this topology is also not used now.

3.3. Buck-Boost Converter

W. Lei, L. Hongpeng, J. Shigong and X. Dianguo [6] in the year 2008 proposed a scheme with buck-boost converter fed BLDC . According to the above analysis, Switches S_1 and S_3 should have a symmetrical blocking voltage characteristic. So, the RB-IGBT (Reverse Blocking IGBT) is used. It can block both forward and reverse voltage during its off state. Comparing IGBT with a series connected diode, elimination of the series diode helps to reduce losses by decreasing the on-state voltage across the switching element. Comparing with bridge buck boost PFC converter bridgeless buck-boost PFC converter has one more switch and capacitor, two less slow diodes. However, comparing the conduction path of these two circuits, at every moment, three semiconductor devices are only conducting for bridgeless buck-boost PFC converter, but four semiconductors are conducting for bridge buck-boost PFC converter. Therefore, conduction loss can be reduced, especially in low line voltage.

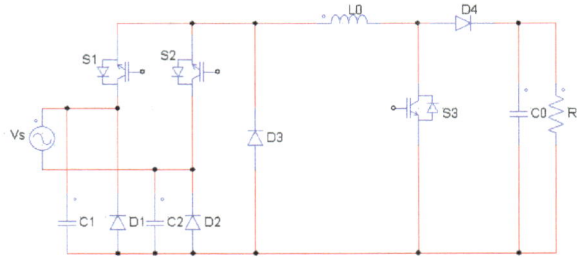

Figure 4. Bridgeless PFC Buck-Boost converter

The above PFC buck-boost converter uses three switches which is cost effective and also increases the switching losses. Thus this method of power quality improvement also has some limitations. So we go for some other topology for better power quality.

3.4. SEPIC PFC Rectifier

A.A. Fardoun, E.H. Ismail, A.J. Sabzali and M.A. Al-Saffar [7] in the year 2012 proposed a method of SEPIC PFC rectifier for BLDC. Figure 5 shows the power stage of a bridgeless SEPIC PFC rectifier. In this circuit, the SEPIC converter is combined with the input rectifier and operates like a conventional SEPIC PFC converter. The operation of this converter is symmetrical in two half-line cycles of input voltage. Therefore, the converter operation is explained during one switching period in the positive half-line cycle of the input voltage. It is assumed that the converter operates in DCM. It means that the output diode turns off before the main switch is turned on. In order to simplify the analysis, it is supposed that the converter is operating at a steady state, and all circuit elements are ideal. In addition, the output capacitance is assumed sufficiently large to be considered as an ideal dc voltage source (V_0). Also, the input voltage is assumed constant and equal to Vac (t_0) in a switching cycle. Based on the aforementioned assumptions, the circuit operation in a switching cycle can be divided into three modes.

The circuit diagram gives the Single Ended Primary Inductance Converter (SEPIC) converter fed BLDC motor for the improvement in the power quality.

Thus the SEPIC converter is efficient but it requires large number of components. So it is not cost effective.

These are some of the bridgeless PFC converter techniques for the improvement of power quality in the ac mains. But all these techniques have some limitations. They also cannot be used for low power applications. So the proposed technique below is designed in such a way that is best suited for low power applications.

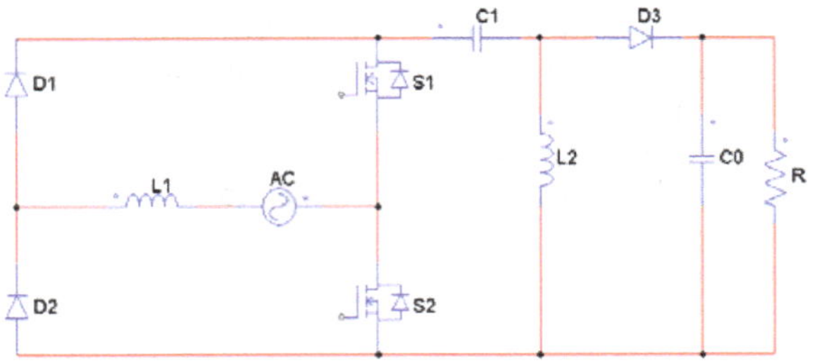

Figure 5. Bridgeless SEPIC PFC converter

4. PROPOSED TOPOLOGY

Figure 6 shows the proposed BL buck–boost converter-based VSI-fed BLDC motor drives. The parameters of the BL buck–boost converter are designed such that it operates in discontinuous inductor current mode (DICM) to achieve an inherent power factor correction at ac mains. The speed control of BLDC motor is achieved by the dc link voltage control of VSI using a BL buck–boost converter. This reduces the switching losses in VSI due to the low frequency operation of VSI for the electronic commutation of the BLDC motor. The performance of the proposed drive is evaluated for a wide range of speed control with improved power quality at ac mains. Moreover, the effect of supply voltage variation at universal ac mains is also studied to demonstrate the performance of the drive in practical supply conditions. Voltage and current stresses on the PFC converter switch are also evaluated for determining the switch rating and heat sink design.

Finally, a software implementation of the proposed BLDC motor drive is carried out to demonstrate the feasibility of the proposed drive over a wide range of speed control with improved power quality at ac mains.

The proposed circuit diagram of the Buck-boost converter fed BLDC motor is shown in the Figure 6.

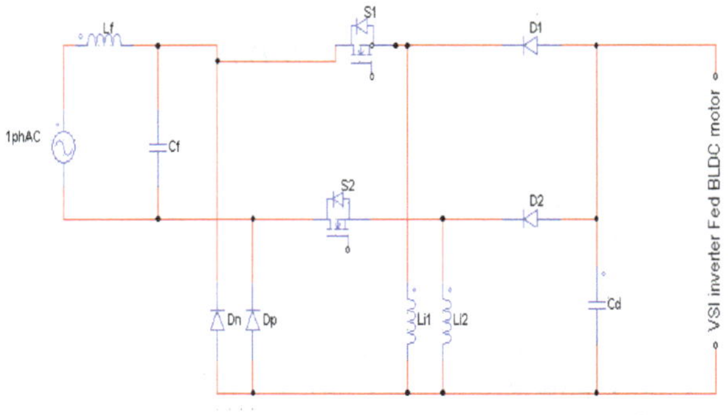

Figure 6. Proposed Circuit diagram of the Buck-boost converter fed BLDC

The above circuit is perfectly suitable for the low power applications.

5. OPERATING PRINCIPLE OF THE PROPOSED PFC BL BUCK– BOOST CONVERTER

The operation of the PFC BL buck–boost converter is classified into two parts which include the operation during the positive and negative half cycles of supply voltage and during the complete switching cycle.

5.1. Operation during Positive and Negative Half Cycles of Supply Voltage

In the proposed scheme of the BL buck–boost converter, switches S_1 and S_2 operate for the positive and negative half cycles of the supply voltage, respectively. During the positive half cycle of the supply voltage, switch S_1, inductor L_{i1}, and diodes D_1 and D_p are operated to transfer energy to dc link capacitor C_d as shown in Figure 7(a)–(c).

Similarly, for the negative half cycle of the supply voltage, switch S_2, inductor L_{i2}, and diodes D_2 and D_n conduct as shown in Figure 8(a)–(c). In the discontinuous mode of operation of the BL buck–boost converter, the current in inductor L_i becomes discontinuous for certain duration in a switching period.

5.2. Operation during Complete Switching Cycle

Three modes of operation during a complete switching cycle are discussed for the positive half cycle of supply voltage as shown hereinafter.

Mode I:

In this mode, switch S_1 conducts to charge the inductor L_{i1}; hence, an inductor current i_{li1} increases in this mode as shown in Figure 7(a). Diode D_p completes the input side circuitry, whereas the dc link capacitor C_d is discharged by the VSI-fed BLDC motor.

Mode II:

As shown in Figure 7(b), in this mode of operation, switch S_1 is turned off, and the stored energy in inductor L_{i1} is transferred to dc link capacitor C_d until the inductor is completely discharged. The current in inductor L_{i1} reduces and reaches zero.

Mode III:

In this mode, inductor L_{i1} enters discontinuous conduction, i.e., no energy is left in the inductor; hence, current i_{li1} becomes zero for the rest of the switching period. As shown in Figure 7(c), none of the switch or diode is conducting in this mode, and dc link capacitor C_d supplies energy to the load; hence, voltage V_{dc} across dc link capacitor C_d starts decreasing. The operation is repeated when switch S_1 is turned on again after a complete switching cycle.

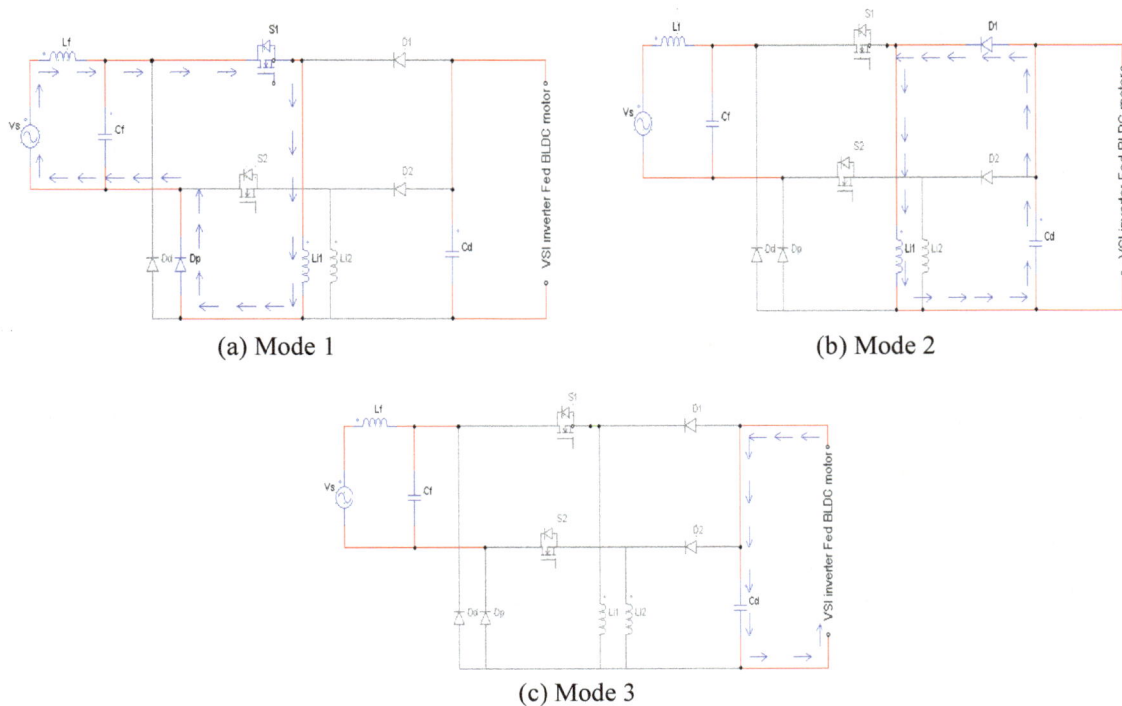

(a) Mode 1   (b) Mode 2

(c) Mode 3

Figure 7. Operation of the proposed converter in different modes (a)-(c) for a positive half cycle of the supply voltage

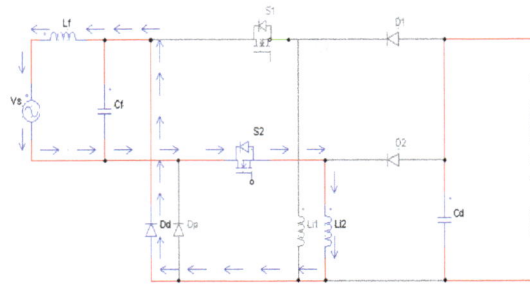

(a) Mode 1

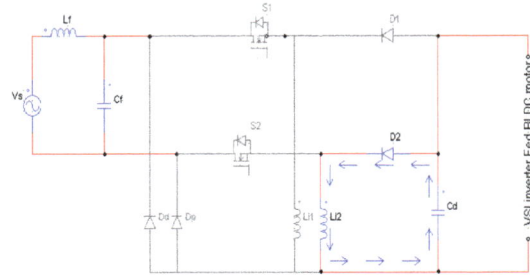

(b) Mode 2

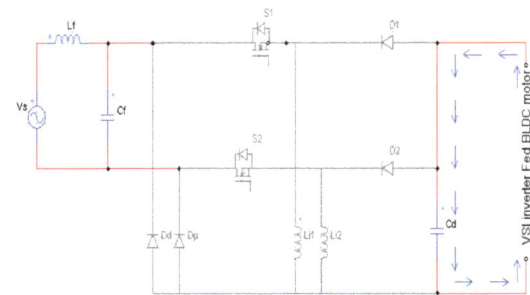

(c) Mode 3

Figure 8. Operation of the proposed converter in different modes (a)-(c) for a negative half cycle of the supply voltage

6. SIMULATION CIRCUIT

The proposed bridgeless buck-boost converter fed BLDC with variable DC link voltage of VSI to improve the power quality at ac mains with reduced components is simulated in MATLAB and the results are shown below.

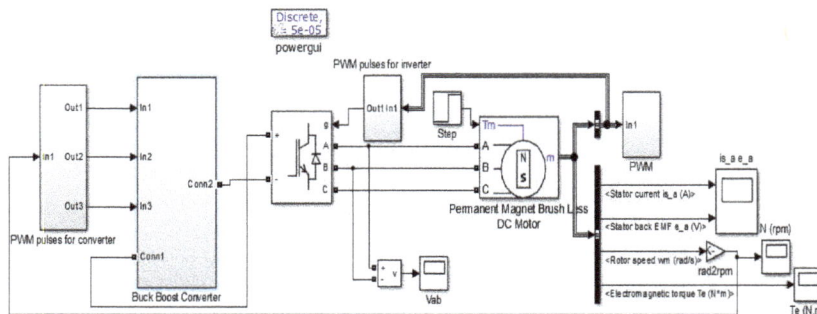

Figure 9. Simulation circuit of Buck-boost converter fed BLDC in MATLAB Simulink

The above circuit consists of the main blocks which are used for the BLDC control. The sub-blocks are presented below. The main sub-block is the buck-boost converter block.

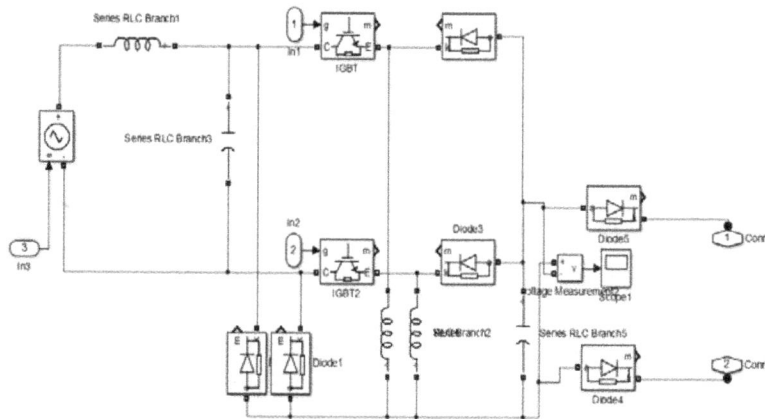

Figure 10. Simulation sub-block of buck-boost converter

The proposed method is simulated in MATLAB as given above and the results are evaluated.

7. SIMULATED RESULTS

The simulated results for various parts of the proposed circuit are shown in Figure 11. The performance of the proposed BLDC motor drive is simulated in MATLAB/Simulink environment using the Sim-Power System toolbox. The performance evaluation of the proposed drive is categorized in terms of the performance of the BLDC motor and BL buck–boost converter and the achieved power quality indices obtained at ac mains. The parameters associated with the BLDC motor such as speed (N), electromagnetic torque (T_e), and stator current (i_a) are analysed for the proper functioning of the BLDC motor. Parameters such as supply voltage (V_s), supply current (i_s), dc link voltage (V_{dc}), inductor's currents (i_{Li1}, i_{Li2},), switch voltages (V_{sw1}, V_{sw2}), and switch currents (i_{sw1}, i_{sw2}) of the PFC BL buck–boost converter are evaluated to demonstrate its proper functioning.

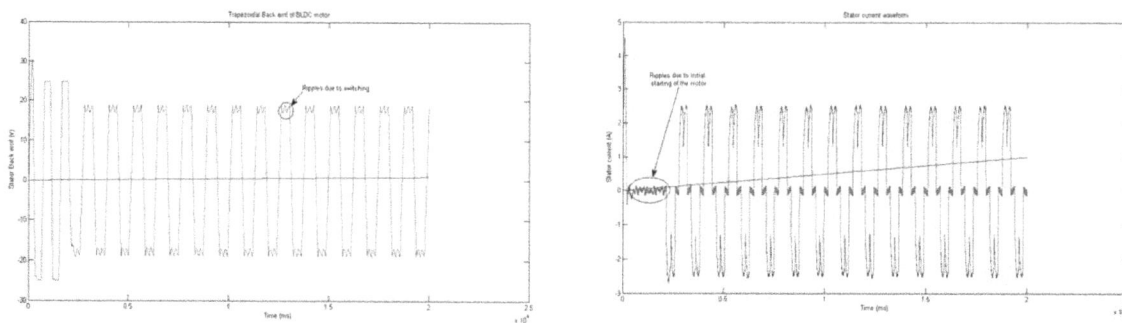

(a) Stator current and electromotive force output waveform

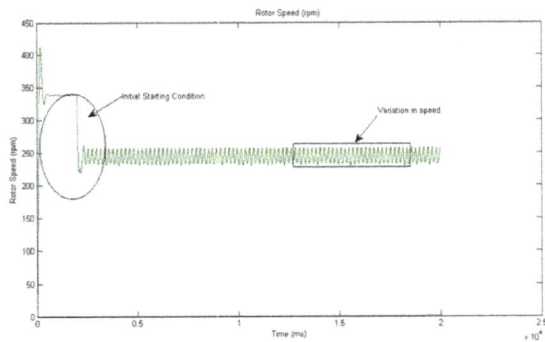

(b) Rotor Speed output waveform

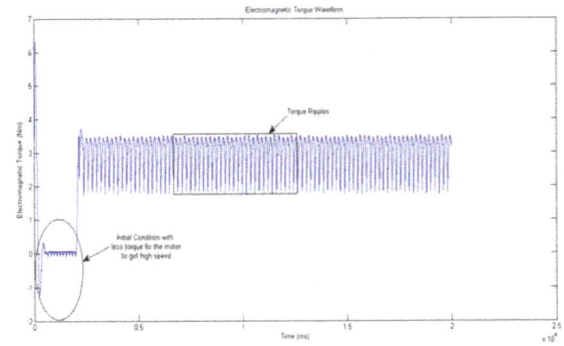

(c) Electromagnetic Torque output waveform

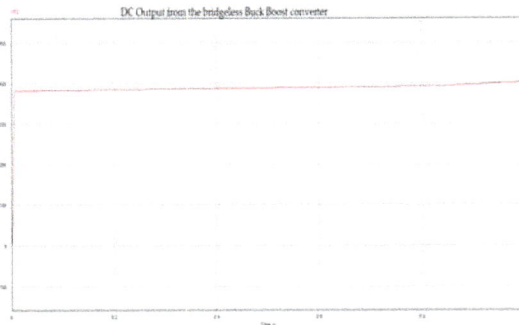

(d) Variable DC voltage output waveform

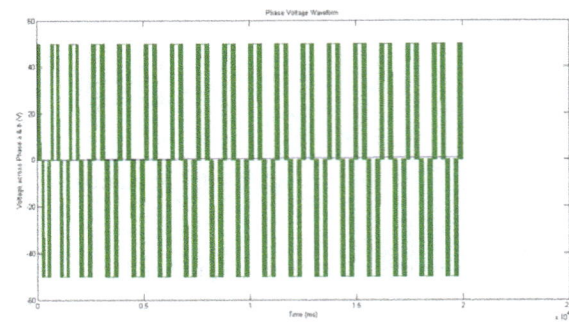

(e) Voltage across phase a & b

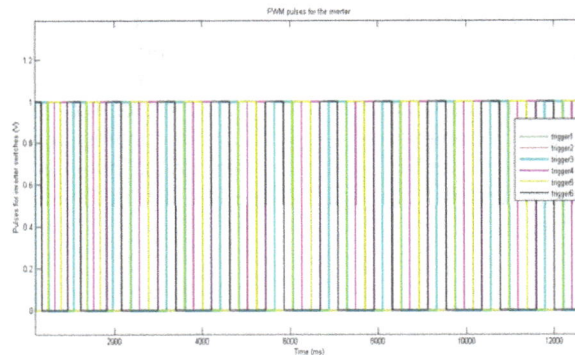

(f) Pulses for VSI

Figure 11. The simulated results for various parts of the proposed circuit

7. CONCLUSION

A PFC BL buck–boost converter-based VSI-fed BLDC motor drive has been proposed targeting low power applications. A new method of speed control has been utilized by controlling the voltage at dc bus and operating the VSI at fundamental frequency for the electronic commutation of the BLDC motor for reducing the switching losses in VSI. The front-end BL buck boost converter has been operated in DICM for achieving an inherent power factor correction at ac mains. A satisfactory performance has been achieved for speed control and supply voltage variation with power quality indices. Moreover, voltage and current stresses on the PFC switch have been evaluated for determining the practical application of the proposed scheme. The proposed scheme has shown satisfactory performance, and it is a recommended solution applicable to low-power BLDC motor drives.

REFERENCES

[1]   T Gopalarathnam, HA Toliyat. A new topology for unipolar brushless dc motor drive with high power factor. *IEEE Trans. Power Electron.*, 2003; 18(6): 1397–1404.

[2]   S Singh, B Singh. *Power quality improved PMBLDCM drive for adjustable speed application with reduced sensor buck-boost PFC converter.* Proc. 4th ICETET. 2011: 180–184.

[3]   S Singh, B Singh. A voltage-controlled PFC Cuk converter based PMBLDCM drive for air conditioners. *IEEE Trans. Ind. Appl.,* 2012; 48(2): 832–838.

[4]   Y Jang, MM Jovanovi´c. Bridgeless high-power-factor buck converter. *IEEE Trans. Power Electron.,* 2011; 26(2): 602–611.

[5]   L Huber, Y Jang, MM Jovanovi´c. Performance evaluation of bridgeless PFC boost rectifiers. *IEEE Trans. Power Electron.,* 2008; 23(3): 1381–1390.

[6]   W Wei, L Hongpeng, J Shigong, X Dianguo. A novel bridgeless buck-boost PFC converter. *IEEE PESC/IEEE Power Electron. Spec. Conf.,* 2008: 1304–1308.

[7]   AA Fardoun, EH Ismail, AJ Sabzali, MA Al-Saffar. New efficient bridgeless Cuk rectifiers for PFC applications. *IEEE Trans. Power Electron.,* 2012; 27(7): 3292–3301.

Permissions

All chapters in this book were first published in IJPEDS, by Institute of Advanced Engineering and Science (IAES); hereby published with permission under the Creative Commons Attribution License or equivalent. Every chapter published in this book has been scrutinized by our experts. Their significance has been extensively debated. The topics covered herein carry significant findings which will fuel the growth of the discipline. They may even be implemented as practical applications or may be referred to as a beginning point for another development.

The contributors of this book come from diverse backgrounds, making this book a truly international effort. This book will bring forth new frontiers with its revolutionizing research information and detailed analysis of the nascent developments around the world.

We would like to thank all the contributing authors for lending their expertise to make the book truly unique. They have played a crucial role in the development of this book. Without their invaluable contributions this book wouldn't have been possible. They have made vital efforts to compile up to date information on the varied aspects of this subject to make this book a valuable addition to the collection of many professionals and students.

This book was conceptualized with the vision of imparting up-to-date information and advanced data in this field. To ensure the same, a matchless editorial board was set up. Every individual on the board went through rigorous rounds of assessment to prove their worth. After which they invested a large part of their time researching and compiling the most relevant data for our readers.

The editorial board has been involved in producing this book since its inception. They have spent rigorous hours researching and exploring the diverse topics which have resulted in the successful publishing of this book. They have passed on their knowledge of decades through this book. To expedite this challenging task, the publisher supported the team at every step. A small team of assistant editors was also appointed to further simplify the editing procedure and attain best results for the readers.

Apart from the editorial board, the designing team has also invested a significant amount of their time in understanding the subject and creating the most relevant covers. They scrutinized every image to scout for the most suitable representation of the subject and create an appropriate cover for the book.

The publishing team has been an ardent support to the editorial, designing and production team. Their endless efforts to recruit the best for this project, has resulted in the accomplishment of this book. They are a veteran in the field of academics and their pool of knowledge is as vast as their experience in printing. Their expertise and guidance has proved useful at every step. Their uncompromising quality standards have made this book an exceptional effort. Their encouragement from time to time has been an inspiration for everyone.

The publisher and the editorial board hope that this book will prove to be a valuable piece of knowledge for researchers, students, practitioners and scholars across the globe.

List of Contributors

M. N. Tandjaoui
University of Bechar, Department of Technology, BP 417 Algeria

C. Benachaiba
University of Bechar, Department of Technology, BP 417 Algeria

O. abdelkhalek
University of Bechar, Department of Technology, BP 417 Algeria

B. Denai
University of Bechar, Department of Technology, BP 417 Algeria

Y. Mouloudi
University of Bechar, Department of Technology, BP 417 Algeria

Gudimetla Ramesh
Departement of Electrical and Electronics Engineering, Jawaharlal Nehru technological university Kakinada

Kari Vasavi
Departement of Electrical and Electronics Engineering, K L University

S. Lakshmi Sirisha
Departement of Electrical and Electronics Engineering, Jawaharlal Nehru technological university Kakinada

M. H. N Talib
Departement of Electrical Engineering, Universiti Teknikal Malaysia Melaka

Z. Ibrahim
Departement of Electrical Engineering, Universiti Teknikal Malaysia Melaka

N. Abd. Rahim
UMPEDAC, Universiti Malaya, Kuala Lumpur, Malaysia

A. S. A. Hasim
Faculty of Engineering, Universiti Pertahanan Nasional Malaysia, Kuala Lumpur, Malay

Rossi Passarella
Department of Computer Engineering, University of Sriwijaya, Palembang, Indonesia

Ahmad Fali Oklilas
Department of Computer Engineering, University of Sriwijaya, Palembang, Indonesia

Tarida Mathilda
Department of Computer Engineering, University of Sriwijaya, Palembang, Indonesia

Prashant Kumar
Departement of Electrical Engineering, Ashokrao Mane Group of Institutes, Shivaji University, Maharshtra, India

Pradip Kumar Sadhu
Electrical Engineering Department, Indian School of Mines (under MHRD, Govt. of India), Dhanbad - 826004, India

Palash Pal
Department of Electrical Engineering, Saroj Mohan Institute of Technology (Degree Engineering Divison), a Unit of Techno India Group, Guptipara, Hooghly-712512, India

Nitai Pal
Electrical Engineering Department, Indian School of Mines (under MHRD, Govt. of India), Dhanbad - 826004, India

Sourish Sanyal
Department of Electronics and Communication Engineering, Academy of Technology, Hooghly, India

Kopella Sai Teja
Departement of Electrical and Electronics Engineering, K L University

R. B. R. prakash
Departement of Electrical and Electronics Engineering, K L University

Jarupula Somlal
Departement of Electrical and Electronics Engineering, K L University, Guntur, INDIA

Venu Gopala Rao.Mannam
Departement of Electrical and Electronics Engineering, K L University, Guntur, INDIA

Narsimha Rao. Vutlapalli
Departement of Electrical and Electronics Engineering, K L University, Guntur, INDIA

Ridwan Gunawan
Department of Electrical Engineering, University of Indonesia, Depok 16424, Indonesia

Feri Yusivar
Department of Electrical Engineering, University of Indonesia, Depok 16424, Indonesia

Budiyanto Yan
Department of Electrical Engineering, University of Muhammadiyah, Jakarta 10510

N. C. Lenin
VIT University, Chennai-600 127, Tamilnadu, India

R. Arumugam
SSN College of Engineering, Chennai -603 110, Tamilnadu, India

M Deepthisree
Departement of Electrical and Electronics Engineering, Amrita Vishwa Vidyapeetham, Amrita School of Engineering, Bengaluru, India

K Ilango
Departement of Electrical and Electronics Engineering, Amrita Vishwa Vidyapeetham, Amrita School of Engineering, Bengaluru, India

V S Kirthika Devi
Departement of Electrical and Electronics Engineering, Amrita Vishwa Vidyapeetham, Amrita School of Engineering, Bengaluru, India

Manjula G Nair
Departement of Electrical and Electronics Engineering, Amrita Vishwa Vidyapeetham, Amrita School of Engineering, Bengaluru, India

Yong-Kun Lu
School of Electronic Information and Automation, Tianjin University of Science and Technology, Tianjin, China

Ade Erawan Minhat
Department of Electronic and Computer Engineering, Universiti Teknologi Malaysia

Nor Hisham Hj Khamis
Department of Communication Engineering, Universiti Teknologi Malaysia

Azli Yahya
Department of Electronic and Computer Engineering, Universiti Teknologi Malaysia

Trias Andromeda
Department of Electronic and Computer Engineering, Universiti Teknologi Malaysia

Kartiko Nugroho
Department of Biotechnology and Medical Engineering, Universiti Teknologi Malaysia

A. V. Kireev
Scientific and Technical Center "PRIVOD-N", Rostov region, Russia

N. M. Kozhemyaka
Scientific and Technical Center "PRIVOD-N", Rostov region, Russia

G. N. Kononov
Scientific and Technical Center "PRIVOD-N", Rostov region, Russia

Hossein Shahinzadeh
Department of Electrical Engineeing, Amirkabir University of Technology, Tehran, Iran

Gevork B. Gharehpetian
Department of Electrical Engineeing, Amirkabir University of Technology, Tehran, Iran

S. Hamid Fathi
Department of Electrical Engineeing, Amirkabir University of Technology, Tehran, Iran

Sayed Mohsen Nasr-Azadani
Department of Electrical Engineeing, Amirkabir University of Technology, Tehran, Iran

Deepak Kumar
Departement of Electrical Engineering, National Institute of Technology, Hamirpur (HP), India

Zakir Husain
Departement of Electrical Engineering, National Institute of Technology, Hamirpur (HP), India

Elakhdar Benyoussef
Faculty of Science and Engineering, Department of Electrical Engineering, University of Djilali Liabes, Sidi Bel Abbes 22000, BP 89 Algeria, Intelligent Control Electronic Power System laboratory (I.C.E.P.S)

Abdelkader Meroufel
Faculty of Science and Engineering, Department of Electrical Engineering, University of Djilali Liabes, Sidi Bel Abbes 22000, BP 89 Algeria, Intelligent Control Electronic Power System laboratory (I.C.E.P.S)

Said Barkat
Faculty of Technology, Department of Electrical Engineering, University of M'sila, Ichbilia Street, M'sila 28000, BP 166 Algeria

S Mohan Krishna
School of Electrical Engineering, VIT University, Chennai, India

J. L Febin Daya
School of Electrical Engineering, VIT University, Chennai, India

Benheniche Abdelhak
Département d'Electrotechnique, Université Badji Mokhtar, BP.12 Annaba, 23000, Algérie

Bensaker Bachir
Laboratoire des Systèmes Electromécaniques, Université Badji Mokhtar, BP.12, Annaba, 23000, Algérie

N.V. Grebennikov
Rostov State Transport University, Rostov-on-Don, Russia

A.V. Kireev
Science and Technology Center "PRIVOD-N", Rostov-on-Don, Russia

C. Nagamani
Research Scholar, University College of Engineering, Osmania University, Hyderabad, India

R. Somanatham
HOD, Dept. Of Electrical & Electronics Engineering, Anurag College of Engineering, Hyderabad, India

U. Chaitanya Kumar
M.Tech Student, Dept. Of EEE, Anurag College of Engineering, Hyderabad, India

A. Petrushin
Chair "Electric Rolling Stock", Rostov State Transport University, Rostov-on-Don, Russia

M. Tchavychalov
Chair "Electric Rolling Stock", Rostov State Transport University, Rostov-on-Don, Russia

E. Miroshnichenko
Chair "Electric Rolling Stock", Rostov State Transport University, Rostov-on-Don, Russia

A. Jeya Selvan Renius
Department of Electrical and Electronics Engineering, School of Electrical Sciences, Karunya Institute of Technology & Sciences University, Coimbatore – 641114, Tamilnadu, India

K. Vinoth Kumar
Department of Electrical and Electronics Engineering, School of Electrical Sciences, Karunya Institute of Technology & Sciences University, Coimbatore – 641114, Tamilnadu, India